Photochemistry and Polymeric Systems

Photochemistry and Polymeric Systems

Edited by

John M. Kelly
Trinity College, Dublin

Ciaran B. McArdle
Loctite International, Dublin

Michael J. de F. Maunder
Speedibrews, Woking

ROYAL SOCIETY OF CHEMISTRY

The Proceedings of a Symposium entitled 'Light on Polymers – Photochemistry and Advanced Materials' held by the Faraday and Industrial Divisions of the Royal Society of Chemistry at Trinity College, Dublin, Ireland, on 16–18 September 1992.

Special Publication No. 125

ISBN 0-85186-485-6

A catalogue record for this book is available from the British Library

© The Royal Society of Chemistry 1993

All Rights Reserved
No part of this book may be reproduced or transmitted in any form or by any means—graphic, electronic, including photocopying, recording, taping or information storate and retrival systems—without written permission from The Royal Society of Chemistry

Published by The Royal Society of Chemistry,
Thomas Graham House, Science Park, Cambridge
CB4 4WF

Printed in Great Britain by Bookcraft (Bath) Ltd.

Preface

The papers collected in this volume were given at the Faraday and Industrial Divisions' Symposium "Light on Polymers - Photochemistry and Advanced Materials" and the Industrial Division's half-day symposium on Photography at the Royal Society of Chemistry's Autumn meeting held in Dublin in September 1992. The organisers' aim was to bring together a number of leading researchers, both chemists and physicists, from academia and industry, in order to consider the fundamental scientific principles and the technological applications of the physical and chemical effects of light on monomers, polymers and polymer-based composites. We have grouped the papers under the following main headings; Radiation curing; Photoablation; Photoluminescence of Polymers; Electro-optics and Nonlinear optics; Photochromism and Photoimaging; Photography. However a primary objective of such an interdisciplinary meeting was to highlight links between these topics and we trust that this is still apparent in these proceedings.

The last two decades has seen dramatic advances in the application of photochemistry to materials' preparation (e.g. for integrated circuits, surface modification, data storage). Such developments and prospects for the future are reviewed by Ledwith (Pilkington,UK), and illustrated by applications in digital optical recording, fibre optics based chemical sensors, stereolithography, and holography. Most such systems involve radiation curing of acrylates using free radical initiators. The development of new photo-initiators (e.g. for work at short u.v. wavelengths) and a detailed mechanistic study with the well-known Irgacure^R 907 initiator using photo-DSC and CIDNP are discussed by Davidson (University of Kent). Other methods for monitoring photocuring processes of acrylates are described in the papers by Decker (CNRS Mulhouse) (using pulsed laser methods with UV or IR monitoring) and by Pethrick (University of Strathclyde) (using viscosity/gel time measurements). Decker also describes the properties of new high performance acrylate monomers.

Another approach to polymeric materials processing is the use of high intensity u.v. lasers to photoablate the materials. In his review of this topic, Srinivasan (UV Tech, U.S.A.), who pioneered much of the work in this area, details some methods for rapidly probing the process. As he points out there is still a lot to learn about the primary processes of photoablation. Of particular interest is whether the ablation is caused by (multiphoton) photochemistry or by thermolysis due to the uv-light-generated heat. A detailed study of the photoablation of polyetheretherketone (PEEK) and polyethyleneterephthalate (PET) is given by Oldershaw (University of Hull) . It is shown that, at least for XeCl laser photoablation of PET, the photothermal mechanism is most important. An interesting application of photoablation [reported by Blau (Trinity College Dublin)] is the formation of waveguides in conjugated polymers.

Photophysical methods have been used for many years to probe the properties of polymers and the continually improving methodology of time-resolved fluorescence methods continues to provide new insights about molecular order and motion as discussed by Phillips (Imperial College London) in his article. Applications to the photoresist polyester PPDA, vinyl aromatic polymers and to surface effects in polydiacetylene are reported. Pethrick (University of Strathclyde) discusses the use of fluorescence methods to probe the nature of the structures responsible for the formation of gels and crystals in isotactic polystyrene.

The optimisation of optical nonlinearities in polymeric systems continues to be an area for close collaboration involving physicists, chemists and electronics specialists. Williams (Eastman Kodak,USA) discusses the design and characterisation of polymers for second order nonlinear optical (NLO) applications. Orientational and relaxational effect in poled polymers containing push-pull chromophores (such as 4-amino-4'-sulphone-azobenzenes) have been studied, using a technique which simultaneously measures second harmonic signal intensity and thermally stimulated discharge current. The same group also report that Langmuir-Blodgett films from suitably derivatised polymers containing similar chromophores exhibit large second order NLO susceptibilities. Blau (Trinity College Dublin) has concentrated on third order NLO phenomena in conjugated polymers such as polydiacetylenes and polythiophenes. Using laser induced grating methods it is shown that these materials have switching times in the picosecond range, which suggest their application in all-optical switching devices, although stability problems still remain to be addressed. Cross et al. (University of Durham) have also studied third order NLO processes but in polymer matrices containing quantum-confined microcrystals of 1:1 TCNQ salts. The convenience of the thin film format of these samples suggests their use for waveguiding. The research reported by Bradley (University of Cambridge) is a striking example of how device physics and macromolecular engineering have been combined. He describes work on the electro-luminescence of poly(p-phenylene vinylene) [PPV], where applications as light emitting diodes and large flat panel displays can be envisaged. Colour tuning has been achieved by judicious choice of PPV-copolymers with appropriate energy gaps, and devices can now be operated with threshold voltages as low as 4V and at 2% efficiency.

The ability to reversibly switch between differently coloured states makes photochromics of interest for a number of applications ranging from digital devices to sunglasses. Crano (PPG Industries, USA) summarises the results of a long-range research project, aimed at the production of plastic organic photochromic prescription eyewear. This necessitated the development of fast-reversing, photochemically-sensitive spiroxazines with long-term stability. Heller (Univerity of Cardiff) describes the preparation of fatigue-resistant photochromic fulgides for digital optical recording. In particular his group, by suitable molecular engineering approaches, has developed systems operating at near infra-red laser diode wavelengths. Crowded helical fulgides, which exhibit optical activity and hence offer the possibility of polarisation-selective readout, are also reported. Irie (Kyushu University, Japan) has addressed some similar problems but from a different perspective and with heteroatomic diarylethenes. Long wavlength (680 nm) absorbing species with excellent fatigue resistance have been prepared and non-destructive optical readout has been achieved using a thermal threshold for the photochromic reaction in conjunction with an optically-induced thermal-biasing technique. A quite different approach to photochromic systems is that described by Hampp exploiting the photochemical properties of mutated bacteriorhodopsins (BR). Genetic, as opposed to molecular, manipulation has been used to produce customised BR variants with optimised polarisation, absorption, refractive index and excited state lifetime

Preface

properties. Two applications already demonstrated for these biomaterials are in holography and real-time optical correlators for pattern recognition.

Fatigue in organic layered photoconductors, which is responsible for the diminished performance of most photocopiers, is an interesting problem in applied photochemistry described by Pacansky (IBM, Almaden, USA). Two major reasons for light-induced fatigue were identified. First the photochemical isomerisation of the hydrazone which is the active component of the charge transport layer. Computational methods were used to assist in designing molecules so as to minimise the photoisomerisation but maintain high efficiency hole transfer. The second major photochemical reaction was due to photo-induced electron transfer between the hydrazone and chlorodiazine blue, a component of the charge generating layer. In this case also computational methods proved invaluable in helping to elucidate the mechanism.

The four papers on Photographic Chemistry illustrate different aspects of the subject. Sahyun (3M, USA) discusses recent advances in our understanding of photographic development. The kinetics of growth of the silver(0) from the latent image can now be mathematically modelled, which should prove very useful for the design of improved photographic emulsions. Fyson (Kodak, UK) describes the changes in chemistry required to reduce the processing time for colour negative films from one hour to 7.5 minutes. This was achieved principally by working at higher temperatures - a course that was only possible by raising the "aqueous melting point" of the gelatin. The papers by Ware (University of Manchester) and Patel (3M, UK) deal with alternatives to silver-based processes. Ware reviews the history and chemistry of the photographic printing methods based on iron polycarboxylates and platinum, palladium or gold. The prints so produced can be of exceptionally high quality. For many purposes photographic processes not involving noble metals are of interest, despite their generally much lower sensitivity than the silver process. Patel outlines methods based on photo-induced electron transfer between oxonol dyes and diphenyliodonium salts. Using a combination of 4-oxonols good colour separation is obtained yielding full colour images.

Contents

Photochemistry and Materials: Progress and Opportunity 1
    A. Ledwith

Radiation Curing

Some New Developments in Radiation Curing 15
    N. Arsu, R. Bowser, R.S. Davidson, N. Kahn, P.M. Moran, and C.J. Rhodes

Photo-induced Polymerization - Kinetic Study of Ultrafast Processes Induced by UV Radiation or Lasers 32
    C. Decker

Photoablation

Interactions of Polymer Surfaces with Ultraviolet Laser Pulses 47
    R. Srinivasan

Excimer and CO_2 Laser Ablation of Organic Polymers 54
    G.A. Oldershaw

Photoluminescence of Polymers

Photophysical Studies of Gelation and Cure in Polymeric Systems 68
    R.A. Pethrick, B. Wandelt, D.J.S. Birch, R.E. Imhof, and S. Radhakrishnan

Luminescence in Synthetic Polymers 87
    D. Phillips

Electro-optics and Nonlinear Optics

Third-order Nonlinear Optical Processes in Conjugated Polymers - From Materials to Waveguide Devices 102
    W.J. Blau

Polymer/CT Salt Composites as $\chi^{(3)}$ Media 110
    G.H. Cross, M. Carroll, T.L. Axon, D. Bloor, R. Rangel-Rojo, A.K. Kar, and B.S. Wherrett

Conjugated Polymer Electro-optic Devices 120
    D.D.C. Bradley, A.R. Brown, P.L. Burn, J.H. Burroughes, R.H. Friend, N.C. Greenham, R.W. Gymer, A.B. Holmes, A.M. Kraft, and R.N. Marks

The Design and Characterization of Polymers for Second-order Nonlinear Optics 134
    D.J. Williams

Photochromism and Photoimaging

Mutated Bacteriorhodopsins - New Biopolymers for Optical Information Processing ... 145
   N. Hampp

The Design and Development of Photochromic Systems for Commercial Applications ... 156
   H.G. Heller, C.C. Elliot, K. Koh, S. Al-Shihry, and J. Whittall

Photochromic Polymers for Optical Data Storage Media ... 169
   M. Irie

Photochromic Organic Compounds in Polymer Matrices ... 179
   J.C. Crano, C.N. Welch, B. Van Gemert, D. Knowles, and B.Anderson

Experimental and Theoretical Studies on the Electrical Fatigue of Organic Photoconductors ... 194
   J. Pacansky, R.J. Waltman, R.J. Berry, and J. Carter

Photography

The Chemistry and Applications of Oxonol-Iodonium Salts ... 209
   R.C. Patel and A.W. Mott

Recent Developments in Photographic Development ... 222
   M.R.V. Sahyun

Faster and Faster Colour Processing ... 236
   J.R. Fyson

Noble Metals for Common Images ... 252
   M.J. Ware

Subject Index ... 267

Photochemistry and Materials: Progress and Opportunity

A. Ledwith

PILKINGTON PLC, LATHOM, ORMSKIRK, LANCASHIRE L40 5UF, UK

During the past two decades there has been a huge increase in the nature and reaction type, wavelengths of light employed, and practical applications of photochemically induced molecular transformations leading to enhancement of both specific and general material properties.

Established commercial technologies include lithography for integrated circuit manufacture[1], from which the entire field continues to derive much stimulus, and equally important ultra violet/visible light curing processes in which wholly reactive liquid or solid precursors are converted to appropriate solids with designed mechanical, chemical and surface properties[2].

For both lithography and more general applications of radiation curing there is a choice between radiation induced molecular step transformations (i.e. one chemical transformation per photon utilised) or radiation induced chain polymerisations which may proceed via free radical or cationic propagating species. Examples include printing and all types of surface coatings, sealants and adhesives, laminates, dental treatments and three dimensional solid imaging. More recent developments utilising rather similar photochemical and molecular principles include holographic imaging[3], surface modification procedures[4], alignment in liquid crystal displays[5], variable light transmission[6], and photovoltaic energy sources[7].

A number of these areas are covered by the proceedings of the Symposium (reported elsewhere in this volume) and the purpose of this presentation is to provide a general perspective on major developments which have occurred over the last twenty years or so and to provide an introduction to the more detailed discussions of specific aspects which follow.

Photoinitiation (Photocure) by Free Radical Processes

In free radical photocure systems the overall requirement is to initiate free radical polymerisation and/or crosslinking by a light induced process. Leaving aside the rather obvious complications which could arise from internal conversions between photochemically excited singlet and triplet states and the possibilities of quenching (including oxygen quenching) and sensitisation, there are essentially three major ways of generating free radical intermediates from a photochemically excited species:

1. Homolytic Cleavage

$$A - B^* \longrightarrow A\cdot + B\cdot$$

2. Hydrogen Abstraction

$$A^* + RH \longrightarrow AH\cdot + R\cdot$$

3. Electron Transfer

(a) $A^* + DH \longrightarrow (A^{\cdot -} DH^{\cdot +}) \longrightarrow AH\cdot + D\cdot$

(b) $(A^+)^* + X^- \longrightarrow A\cdot + X\cdot$

Homolytic cleavage leading to formation of free radicals has been known longest and is manifest in the use of benzoin and its alkyl ethers in photoinitiation. Related molecules such as hydroxyalkylketones, dialkoxyacetophenones and benzoylphosphine oxides[8] are also utilised and in every case, simple alpha cleavage generates a benzoyl radical and a more reactive radical derived from the appropriate molecular skeleton.

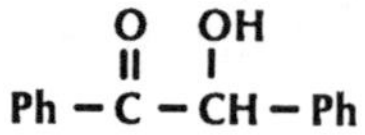

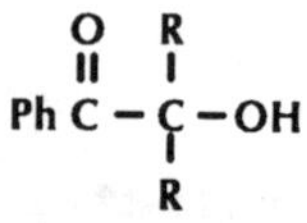

$$\underset{\text{benzoin}}{Ph-\overset{O}{\overset{\|}{C}}-\overset{OH}{\underset{}{C}H}-Ph} \qquad \underset{\text{benzoin alkyl ether}}{Ph-\overset{O}{\overset{\|}{C}}-\overset{OR}{\underset{}{C}H}-Ph} \qquad \underset{\text{hydroxyalkyl ketones}}{Ph\,\overset{O}{\overset{\|}{C}}-\overset{R}{\underset{R}{C}}-OH}$$

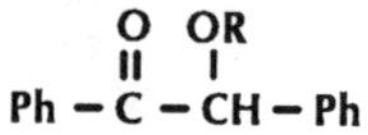

$$\underset{\text{dialkoxyacetophenones}}{Ph\,\overset{O}{\overset{\|}{C}}-\overset{OR}{\underset{OR}{C}}-R^1} \qquad \underset{\text{benzoylphosphine oxides}}{Ph\,\overset{O}{\overset{\|}{C}}-\overset{O}{\overset{\|}{P}}\!\!<\!\!\overset{R}{\underset{R^1}{}}}$$

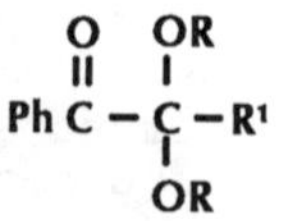

A useful survey of quantitative aspects of the use of u.v. initiators, and some indication of newer molecular structures, has been given by Fouassier[9].

Hydrogen abstraction from organic molecules by photochemically excited ketones such as benzophenone is one of the best known reactions in molecular photochemistry. However it is not commonly employed for photoinitiation except when used in conjunction with an electron donor such as an amine or an amino-substituted benzophenone like Michlers Ketone. Electron transfer processes are prevalent in reactions between donor molecules (such as amines) and aromatic ketones which have lowest lying singlet states with π,π^* character. Intersystem crossing to reactive n,π^* triplet states then provides a mechanism for radical formation. An additional feature of such systems is the relatively strong absorption at wavelengths greater than 400 nm which facilitates light induced curing in pigmented systems. Typical examples of ketones participating in photoinduced electron transfer processes are indicated below.

$(CH_3)_2N$-⌬-$\overset{O}{\overset{\|}{C}}$-⌬-$N(CH_3)_2$ $Ph\overset{O}{\overset{\|}{C}} - \overset{O}{\overset{\|}{C}}Ph$

Michlers ketone   **benzil**

2 - chlorothioxanthone   **camphorquinone**

Whilst many tertiary amine-type donor molecules may be employed with these and related carbonyl compounds, a particularly convenient amine donor class is that comprising the alkyl esters of 4-dimethylamino benzoic acid the reactions of which may be taken as representative of many ketone/amine combinations.

$\overset{|}{\underset{|}{C}} = O + (CH_3)_2N$-⌬-$COOR \xrightarrow{h\nu} \cdot \overset{|}{\underset{|}{C}} - OH + \overset{\cdot}{C}H_2 \overset{|}{\underset{CH_3}{N}}$-⌬-$COOR$

monomer

$HOC - CH_2 - \overset{|}{\underset{CH_3}{N}}$-⌬-$COOR$ $\overset{|}{HOC} - \overset{|}{COH} + ROOC$-⌬-$\overset{|}{\underset{CH_3}{N}} - CH_2$ *polymer*

There are abundant examples of use of many of these ketone and ketone/amine systems in u.v./visible light cure but it is perhaps worth noting the rapid growth in the use of visible light (i.e. wavelengths longer than 400 nm) as a convenient cure mechanism for a variety of dental repairs[10]. Camphorquinone has appropriate light absorption properties and is frequently employed as the ketone component. Many amines may be used but it is especially convenient to include (2-dimethylaminoethyl) methacrylate in the particle filled acrylic resin system. Excellent wear and cosmetic properties are obtained and highly convenient and efficiently focused light sources have been especially designed for the dentist and patient's convenience. Polymerisable single paste, composite, dental cement fillings have been under development and application for a considerable time and use of visible light curing affords maximum convenience - not least because the dentist can stop the cure at any point by removing the light source. This provides for optimised manipulation of the composite paste cement in order to ensure complete, and relatively defect free, filling of cavities and other features as well as effective bonding to the cavity walls.

Photoinitiation by Cationic Processes

Accompanying the widely diverging technologies involving photochemically induced free radical polymerisation, there has been a concurrent awakening of interest in the rather more limited applicability of photochemically induced cationic polymerisations. The potential value of this type of photocure was first elaborated by Crivello[11] and a wide ranging review of appropriate kinetics and mechanisms has been published very recently[12]. Important aspects are covered by Davidson in these proceedings. In outline the process requires direct excitation (u.v. source) or sensitised decomposition (u.v. or visible light) of di-aryliodonium or tri-arylsulphonium salts. Detailed photodecomposition mechanisms vary with the particular system employed but it is widely assumed that formation of protonic acid is an indirect product of photolysis. Provided that the counter ion of the initial onium salt is relatively stable (i.e. non-nucleophilic), the resultant protonic acid will be a good initiator for cationic polymerisation of epoxides and alkyl vinyl ethers e.g.

$$Ph_2I^+ X^- \xrightarrow{h\nu} (Ph_2I^+)^* X^- \begin{cases} \rightarrow PhI + Ph^+ + X^- \\ \rightarrow Ph\cdot + PhI^{\cdot+} + X^- \end{cases}$$

$$Ph^+ + PhI \rightarrow Ph\text{-}PhI + H^+$$

$$H^+ (\text{or } Ph^+) + monomer \rightarrow polymer$$

$$X^- + SbF_6^-, PF_6^- \text{ etc.}$$

The examples noted below serve only to illustrate the scope of current commercially important materials technologies which may be derived from photocure systems or from the basic photochemistry associated with the corresponding photoinitiation process.

Digital Optical Recording and Laser Vision Video Discs

An important application of light curing technology, developed by the Philips Company, is required in the manufacturing process for optical data storage (archival) and the more generally applicable video disc recording media[13]. The process is commonly referred to as the '2p process' (from photopolymerisation) and is based on u.v. light initiated curing of multifunctional acrylic monomers, the development and performance of which has been described by Kloosterboer[14].

2p Process (Philips)

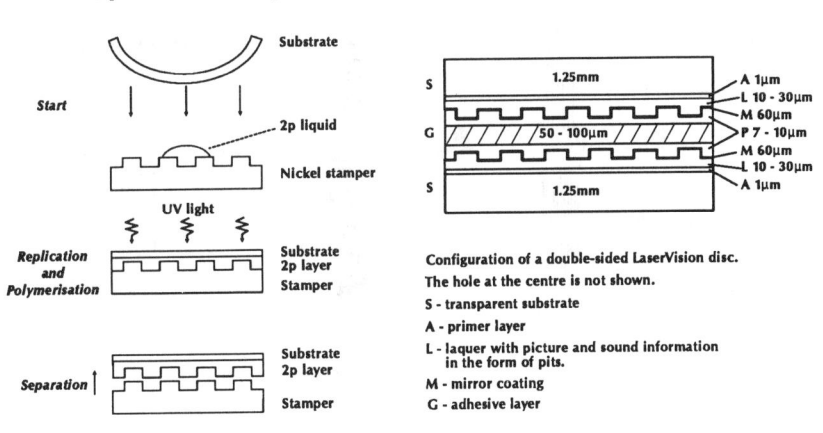

Configuration of a double-sided LaserVision disc.
The hole at the centre is not shown.
S - transparent substrate
A - primer layer
L - laquer with picture and sound information in the form of pits.
M - mirror coating
G - adhesive layer

A brief outline of the process for replicating video discs is shown in the figure and, as indicated, the substrate can be any material provided it is transparent to the wavelengths of light required for initiation of cure. For video discs it is normal to use plastic substrates (e.g. polymethylmethacrylate) whereas for archival storage, which requires larger disc sizes, it is commonplace to use glass. The master mould plate is usually made of metal and following the indicated photocuring of liquid monomer between mould and substrate, the formatted 2p layer is given a laser light read-out reflective metal coating and a further protective polymer layer. Typical disc dimensions and organisation of a finished disc are given in the figure and the information is recorded as a spiral track of small pits about 0.12 microns deep and 0.4 microns wide. The length of the pits and the minimum separation vary between 0.5 and 2.0 microns.

Fibre Optic Based Chemical Sensors

An important objective in clinical medicine is the continuous, *in situ* measurement of concentrations of blood solutes such as pH, CO_2 and O_2. Most multi-analyte sensors developed for *in situ* monitoring consist of several sensors aggregated into a bundle and this is not ideal where sensor size is an important factor. A recent technique, shown schematically in the figure, utilises localised photopolymerisation of appropriate dye indicators on the face of an imaging optical fibre[15]. It is convenient, by this means, to locate and interrogate several different units on a compact multi-analyte fibre optic chemical sensor system.

Sensor Fabrication

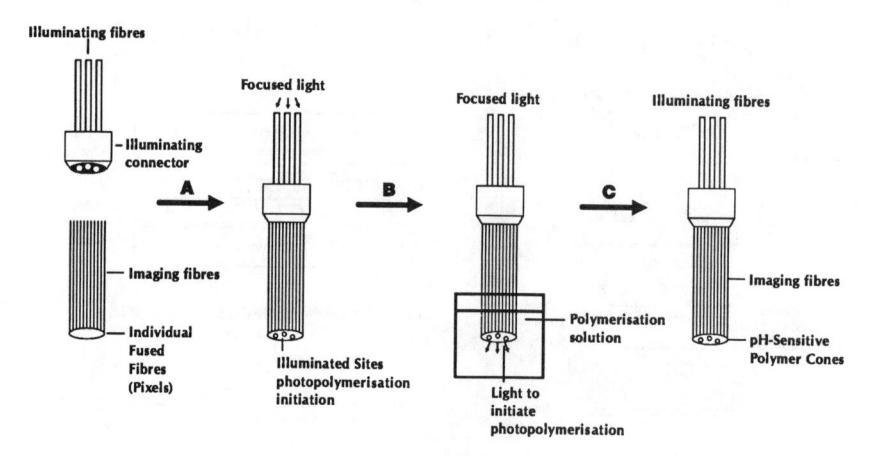

Although the development is at a very early stage, this technique holds great promise and it should be possible to generate micron sized sensing regions. Typically a short length of imaging fibre was attached, by means of a fibre connector, to the distal end of the illuminating fibres which, in turn, direct light to the proximal end of a discrete number of individual pixels that comprise the imaging fibre. It is this readily controlled activation of pixels which determines the areas of the distal face functioning as initiation sites for photopolymerisation of the dye/polymer solution. Illumination fibres are about 125 microns in diameter and the imaging fibres are suspended in a solution of acryloyl fluorescein, 2-hydroxyethylmethacrylate, methylene bisacrylamide and benzoin ethyl ether. The resulting photocure yields cones of polymer entrapped dye at the end of the imaging fibre which may be varied in area and height by adjusting the illuminating light source focus and the time of exposure.

Stereolithography (Photochemical Machining)

Stereolithography is a new technology linking the power of computer graphics to the rapid formation of a solid, shaped object. Patented originally by Hull[16], stereolithography converts CAD/CAM/CAE generated solid or surface model data to a three-dimensional real part synthesised, via photopolymerisation, from a photosensitive monomer such as an acrylate. Cured by a laser beam directed across an x,y surface, a liquid monomer mix is converted to a solid plastic object, point by point, layer by layer, with quality in all three dimensions x, y and z determined by the properties (especially densities) of the monomer/polymer components.

Schematic Representation of the Stereolithography Apparatus

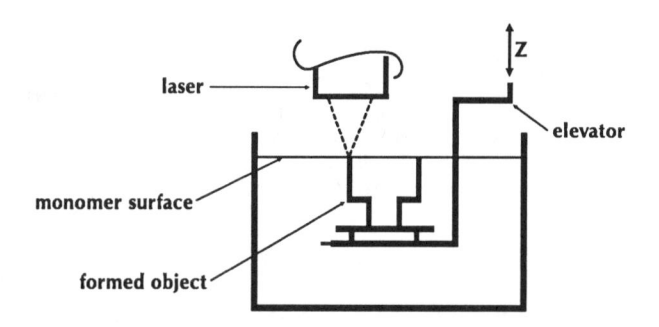

Stereolithography is simple in concept[17] and it provides great economies for the design lab as well as for the modelling process. It also provides previously unrecognised challenges for the polymer photochemist, for it is entirely a laser-initiated technology, and the polymerisation reactions take place to depths below a finitely thin surface layer. Currently, modelled parts made by stereolithography are generally on the order of 3" to 6" on a side. The goal is to make much larger parts and to make them faster and truer. If stereolithography is ever to play a role in manufacturing, parts must be synthesised from widely usable plastics like epoxies and polyurethanes and not just from the rather brittle crosslinking acrylic systems. One of the monomers widely used in stereolithography is bis-phenol-A-diacrylate.

Minimisation of cure shrinkage is one of the greatest challenges in stereolithography resin development. Even the best of the pure acrylates shrink 6-8% in volume when fully converted to polymer from monomer. Most direct laser-initiated photocures are not complete and there always remain unreacted olefinic centres in the polymer. After the laser initiated photopolymerisation there is opportunity for additional radical chemistry to occur in the photopolymer which can change the overall mechanical

properties and shape of the formed polymer. In the best
case today, a maximum deviation of 0.015" over 10" is
achieved. But as part geometry changes, and part sizes
increase, there is even less tolerance for shrink in the
formed photopolymer.

There are many putative no-shrink monomer systems
which may or may not have application in
stereolithography. Among these are the very large number
of ring opening polymerisation systems reported by
Bailey[18] and the thio/norbornene systems reported by
Jacobine[19]. Both are said to minimise shrink.
However stereolithography also requires a rapid
photochemical curing step, and many resins fail
on this count. Clearly, the development of
resin systems amenable to rapid photopolymer formation is
an important step in advancing stereolithography.

For laser induced polymerisation it is necessary to
work with visible light absorbing initiator systems and
these can be free radical or cationic in nature. Lee and
Neckers[20] have recently reported a most interesting two
photon initiator system based on photochromic
spiropyrans. The latter form intensely coloured
merocyanine dyes on exposure to u.v. irradiation which,
in the presence of N-phenylglycine, may then be
selectively activated to produce initiating free radicals
with a helium/neon laser at 632 nm. It is convenient to
use a helium/cadmium laser at 324 nm for the first
photon.

Photochemistry at the intersection of two laser beams is
an obvious way of improving the quality of solid parts
produced by stereolithography.

Holographic Optical Elements

Holographic techniques allow a wide range of devices
to be made from 3 dimensional photographs to specialised
optical elements[3]. An important practical application of
holographic optical elements is as wavelength selective
mirrors for the combiner plates of Head-Up-Displays
(HUDs). In an aircraft head-up-display (HUD),
information is displayed on a cathode ray tube. This can
be navigational aids, weapon aiming markers, or a picture
of the outside terrain produced by forward looking infra

red cameras for flying at night or in bad visibility. Light from the CRT passes through a lens system and is reflected into the pilots line of sight by the holographic HUD combiner element. The pilot views the outside world through the HUD combiner which is located between him and the aircraft window. He therefore sees the CRT display superimposed on his view ahead. The lens system and holographic combiner together produce an image focused virtually at infinity, so the pilot does not need to refocus his eyes to clearly read the display.

Avionic Head Up Display

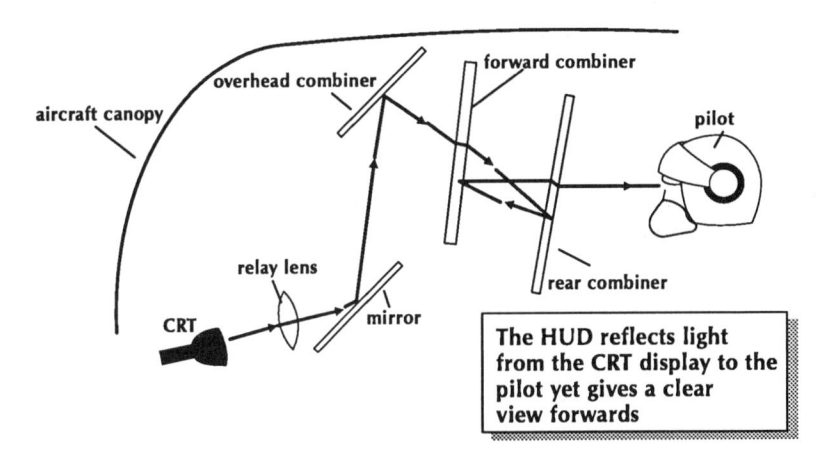

The HUD reflects light from the CRT display to the pilot yet gives a clear view forwards

Usually the holographic combiner has a high reflectivity (nearly 100%) in a narrow band (of around 20 nm) and high transmission outside the reflection band. The hologram is formed in a thin film about 10 to 20 microns thick and usually sealed between two pieces of glass. It comprises a periodic modulation of refractive index in layers spaced about 0.2 microns apart through the film thickness. This microscopic structure operates on light waves by diffraction to filter out a single colour and focus or redirect an image. The wavelength reflected depends on both the angle of reflection and the spacing of the layers. During manufacture this spacing is carefully controlled so the hologram is tuned to reflect the emission band of a chosen narrow band CRT phosphor.

Commercial holograms of this type are most commonly made in a material called dichromated gelatin. Although dichromated gelatin can produce holograms of very high quality and performance, good uniformity and reproducibility of performance can only be achieved with very precise process and environmental control during manufacture.

The dichromated gelatin manufacturing process involves coating a glass substrate with a gelatin film which is made light sensitive by the incorporation of ammonium dichromate. The film is then exposed to a pattern of light and dark optical interference fringes, formed by a laser optical system, designed to produce the required hologram refractive index structure. Where the dichromated gelatin is exposed to a bright fringe, chromium ions are reduced causing localised cross linking of the gelatin molecules. The film is then subjected to a series of processing baths which swell and then shrink the film, and induce a modulated refractive index in the film corresponding to the cross linking pattern and so to the pattern of laser interference fringes. After tuning and stabilising treatments, a glass coverplate is sealed on for protection from the environment. The hologram may be generated either by utilising two independent light beams or, as illustrated, a suitably placed mirror will suffice to produce the necessary interference pattern.

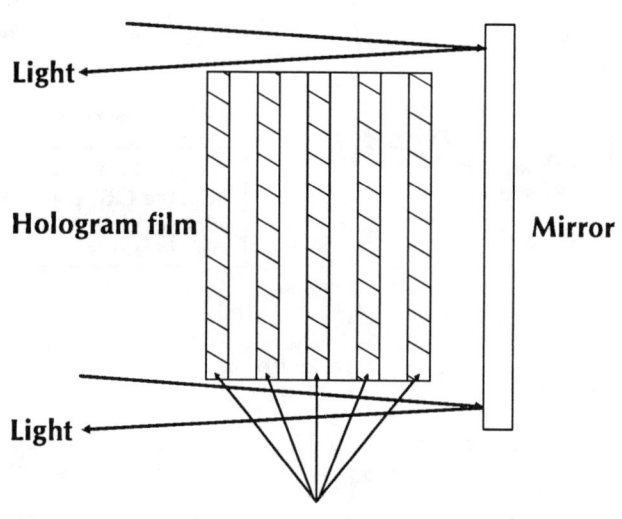

Light

Hologram film

Mirror

Light

Photochemical reaction zones

Disadvantages of gelatin-based holograms lie in the requirement for wet processing and the obvious moisture sensitivity of the final product. More recently the DuPont Company has developed the well known photoinitiating properties of hexa-arylbi-imidazoles (Lophine dimers) into an all dry processable holographic film technology with excellent optical and material characteristics[1,21].

Overall photochemical and chemical transformation are shown below and a wide variety of monomers, binders and plasticisers may be substituted.

BI   Dye   BOSH

Cellulose or Polyvinyl Acetate Binder

Dye $\xrightarrow{h\nu}$ Dye* $\xrightarrow{BI}$ I·

I· + BOSH $\longrightarrow$ BOS· + IH

BOS· $\xrightarrow{\text{Acrylates}}$ Polymer

    Holograms recorded in this way are free of voids or pores and consist of (i) photopolymer-rich regions formed by monomer polymerisation in areas of constructive interference and (ii) binder-rich regions formed by (mainly) monomer diffusion out of areas of destructive interference. The difference in refractive index between monomer and binder controls the refractive index modulation value of the hologram. Typical binders include polyvinylacetate and cellulose acetage butyrate; suitable monomers include 2-phenoxyethylacrylate, halogen substituted phenyl acrylates and solid monomers such as N-vinylcarbazole. Post exposure treatment involves a flood exposure to the light source and heat treatment may be used to improve the monomer diffusion rates.
    The DuPont holographic system is extremely versatile, readily processed and is likely to be very important in the extension of HUD technologies into other areas including road vehicles. It is interesting that a recent Patent[22] describes alternative all dry processable holographic films using similar binders but having a mixture of monomers (e.g. epoxy and acrylic monomers) which may be selectively and independently polymerised by cationic and free radical processes.

Photochemical Surface Modification

    Material science has recently broadened its focus from a sole reliance on physical and mechanical properties. An increasingly important role has been given to material surface characteristics in the design and implementation of product applications. Historically, many devices fail or never reach the market because of incompatible surface attributes. Since no material offers both the structural and interfacial features to meet every need, technologies to modify the surface properties of structurally sound products are needed to help them function efficiently in their unique environments.

A number of photochemical processes may be employed to modify both the physical and chemical properties of surfaces and, as described in these proceedings, laser ablation represents an extreme case. One area of particular importance which is readily addressed by more usual photochemical treatments is associated with modification of synthetic and natural polymer surfaces required for a variety of applications of medical or biological interest.

The PhotoLink™ technology from Bio-Metric Systems, Inc., is an innovative method for surface modification and immobilisation[4]. It offers an easy, versatile method of improving product performance and the photochemistry allows covalent surface coupling to virtually any carbon-containing substrate, including most types of polymers and certain types of glass and metal. Unlike many other coatings, PhotoLink™ reagents can be very thin (on the order of 10 to 1,000 nanometers) and highly uniform.

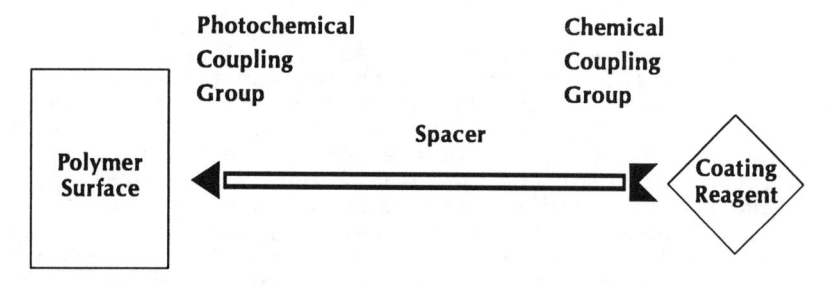

As illustrated, a photochemically active grouping capable of abstraction or insertion reactions with hydrocarbon surfaces, constitutes the basis of the technique. The photoactive group is attached to a spacer which binds the desired coating or surface modifying ingredient. Benzophenones offer the most versatile photo-linking units as illustrated.

The required modification may involve small molecule, oligomeric or polymeric derivatives and typical examples are indicated below.

Substituted Benzophenones

[Chemical structures of substituted benzophenones:

Ph-C(=O)-Ph-C(=O)-(OCH$_2$CH$_2$)$_n$-OH

Ph-C(=O)-Ph-CH$_2$-(OCH$_2$CH$_2$)$_n$-OH

Ph-C(=O)-Ph-C(=O)-NH-CH$_2$-CH$_2$-CH$_2$-NH-C(=O)-C(CH$_3$)(OCH$_2$-CH(NH$_2$))-CH$_3$

Ph-C(=O)-Ph-C(=O)-NH-CH$_2$-CH$_2$-CH$_2$-NH-C(=O)-C(CH$_3$)(OCH$_2$-CH(N-))-CH$_3$-]

Anti-bodies, cells, proteins, peptides and antithrombogenic agents may have their adhesion characteristics altered by this type of surface modification and there are obvious applications in medical devices, cell culture surface and affinity purification systems. More general problems associated with surface wettability are readily addressed by PhotoLink™ techniques.

Future Developments

Space does not permit an exhaustive coverage of the many and diverse opportunities for materials design and modification afforded by photochemical techniques. The constant increase in demands for higher component densities in semi-conductor circuits provides a continuous requirement for design of ever more reliable and processable materials for high resolution photolithography[23]. As noted earlier, this remains a major stimulus to more widely applicable photochemical transformations.

A second major stimulus is provided by the slow but steady increase in utilisation of solar energy as a contribution to both energy conservation and environmental protection. An outstanding example is the recent report by Grätzel of new types of photovoltaic cells derived from organic dye-modified titanium dioxide particles. The future holds much promise and it is certain that there will be a continuous requirement for Symposia such as that represented by these proceedings.

REFERENCES

1. A. Reiser, 'Photoreactive Polymers: The Science and Technology of Resists', Wiley Interscience, N.Y., 1989.
2. C. G. Roffey, 'Photopolymerisation of Surface Coatings', Wiley, 1982; S. P. Pappas, 'U.V. Curing: Science and Technology', Technology Marketing Corp., Norwalk, Connecticut, Vol.1, 1978; Vol.2, 1985.
3. N. Abramson, 'The Making and Evaluation of Holograms', Academic Press, London, 1981.
4. S. G. Dunkirk, S. L. Gregg, L. W. Duran, J. D. Monfils, J. E. Haapala, J. A. Moray, D. L. Clapper, R. A. Amos and P. E. Guire, J. Biomaterials Applications, 1991, 6, 131-156.
5. T. Seki, T. Tamaki, Y. Susuki, Y. Kawanishi, K. Ichimura and K. Aoki, Macromolecules, 1989, 22, 3505; W. M. Gibbons, P. J. Shannon, S. T. Sun and B. J. Swetlin, Nature, 1991, 351, 49.
6. 'Applied Photochromic Polymer Systems', Ed. C. B. McArdle, Blackie and Son Ltd., Glasgow, 1992.
7. B. O'Regan and M. Grätzel, Nature, 1991, 353, 737.
8. T. Sumiyoshi, W. Schnabel, A. Henne and P. Lechtken, Polymer, 1985, 26, 141.
9. J. P. Fouassier, Progress in Organic Coatings, 1990, 18, 229.
10. R. J. Kilian, Chemtech, 1981, 678; see also commissioned report in Journal of American Dental Assoc., 1985, 110, 100.
11. J. V. Crivello and J. H. W. Lam, Macromolecules, 1977, 10, 1307.
12. R. J. DeVoe, P. M. Olofson and M. R. V. Sahyunin, Advances in Photochemistry, Eds. D. Volman, G. Hammond and D. C. Neckers, John Wiley and Sons, 1992, 17, 313.
13. H. C. Haverkorn van Rijsewijk, P. E. J. Legierse and G. E. Thomas, Philips Technical Review, 1982, 40, 287; J. G. Kloosterboer, G. J. M. Lippits and H. C. Meinders, ibid, 1982, 40, 298.
14. J. G. Kloosterboer, Advances in Polymer Science, Springer Verlag, Berlin, 1988, 84, 1-61.
15. S. M. Barnard and D. R. Walt, Nature, 1991, 353, 338.
16. C. Hull, U.S. Patent 4,575,330, 1986.
17. D. C. Neckers, Chemtech, 1990, 615.
18. W. J. Bailey, Polymer J., 1985, 17, 85.
19. A. Jacobine, Polymeric Materials Sci.Eng., 1989, 60, 211.
20. S. K. Lee and D. C. Neckers, Chem.Mater., 1991, 3, 858.
21. W. K. Smothers, B. M. Monroe, A. M. Weber and D. E. Keys, Practical Holography IV, Proc.SPIE 1212, 1990, 20; B. M. Monroe, W. K. Smothers, D. E. Keys, R. R. Krabs, D. J. Mickish, A. F. Harrington, S. R. Schicker, M. K. Armstrong, D. M. T. Chan and C. I. Weather, J.Imaging Sci., 1991, 35, 19.
22. M. Kawabata, A. Sato, I. Sumlyoshi, Europ. Pat. 0,509,512A1, 1992.
23. 'Polymers in Microlithography, Materials and Processes', Eds. E. Reichmanis, S. A. MacDonald and T. Iwayanagi, A.C.S. Symposium Series, Vol.412, 1989.

Some New Developments in Radiation Curing

N. Arsu,[1] R. Bowser,[2] R. S. Davidson,[2*] N. Kahn,[1] P. M. Moran,[1] and C. J. Rhodes[3]

[1] DEPARTMENT OF CHEMISTRY, CITY UNIVERSITY, LONDON EC1V 0HB, UK
[2] THE CHEMICAL LABORATORY, THE UNIVERSITY, CANTERBURY, KENT CT2 7NH, UK
[3] DEPARTMENT OF CHEMISTRY, QUEEN MARY & WESTFIELD COLLEGE, UNIVERSITY OF LONDON, MILE END ROAD, LONDON E1 4NS, UK

SUMMARY

This paper outlines the pros and cons of using allyl/benzylsilanes and stannanes, quinoxalines and some iron-arene complexes as free radical initiators. The chemistry of the Type I photoinitiator 1-(4-methylthiophenyl)-2-methyl-2-(N-morpholino)propan-1-one is discussed.

INTRODUCTION

Over the last five years the radiation curing industry has seen continued growth despite the recession [1]. Undoubtedly the "green" aspects of radiation curing are making it more popular. These are linked to tougher legislation concerning solvent emission, and the level of migratables in food packaging [2]. Another major factor is the increased understanding of the fundamental processes which lie behind radiation curing [3], the development of safer and better prepolymers and new photoinitiators [4], and the development and availability of better equipment. Whilst the greatest tonnage of radiation curable materials is used by the surface coatings industry, a valuable market is the production of resists (for lithographic applications in the printing industry) and of circuits (conventional printed circuit boards down to sub micron sized patterning for production of microchips) [5].

In this article four topics, which are related to our current work will be discussed, namely:

(i)   The photochemistry of allyl/benzylsilanes and stannanes.
(ii)   The use of some iron arene complexes for cationic and free radical polymerisation.
(iii)   Quinoxalines as photoinitiators.
(iv)   Chemistry of the photoinitiator 1-(4-methylthiophenyl)-2-methyl-2-(N-morpholino)propan-1-one (Irgacure^T 907).

RESULTS AND DISCUSSION

(i) **The photochemistry of allyl and benzylsilanes and stannanes.**

In recent work on the electron beam induced curing of silicon and tin containing acrylates [6,7], it was found that these compounds showed high reactivity. This was in part attributed to the greater stopping power of silicon and tin compared with carbon, and also to the possibility that silyl and stannyl radicals were being generated which were acting as efficient initiators for polymerisation. Silyl radicals are known to attack carbon-carbon double bonds with great avidity [8a] and tin radicals will attack carbon-carbon triple bonds [8b]. Since curing by e.b. radiation depends upon ionising substrates to generate radical cations and slow electrons, [3] it was thought that silicon and tin containing polymethacrylates might act as good e.b. resists.

$$eg \quad \overset{}{\underset{CO_2SiR_3}{\wedge\!\!\!\vee\!\!\!\wedge}} \quad \xrightarrow{e.b.} \quad \overset{+\bullet}{\underset{CO_2SiR_3}{\wedge\!\!\!\vee\!\!\!\wedge}} \quad \longrightarrow \quad \underset{CO_2^{\bullet}}{\wedge\!\!\!\vee\!\!\!\wedge} \;+\; R_3Si^{+\bullet}$$

$$\downarrow$$

$$CO_2 \;+\; \text{Chain Scission}$$

For all the materials prepared it was found that they behaved as negative resists thereby suggesting that the main chain was not fragmenting. Clearly further work had to be done to gain a more accurate understanding of the chemistry of silicon and tin radical cations. Interestingly benzylsilicon and tin radical cations have been generated via photoinduced electron transfer reactions [9,10,11] and a number of product studies carried out.

Some New Developments in Radiation Curing

e.g.

(a) [reaction scheme with nBu$_4$M + dicyanobenzene under hv/CH$_3$CN forming radical cation pair, then nBu• + nBu$_3$M, yielding 4-butylbenzonitrile + other products]

M = Si or Sn

Ref 10

(b) Ph-C(=O)-C(=O)-Ph + allyl-SnR$_3$ →(hv) [Ph-C(OSnR$_3$)(allyl)-C(=O)-Ph] → Ph-C(OH)(allyl)-C(=O)-Ph

Ref 11

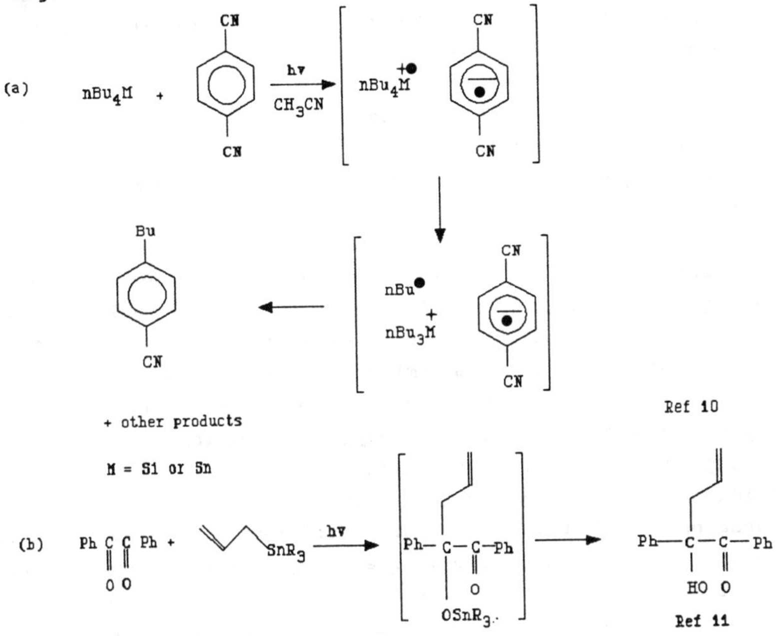

Mariano et al. have utilised, in a most fruitful way, the fragmentation of allylsilane radical cations in synthesis [12].

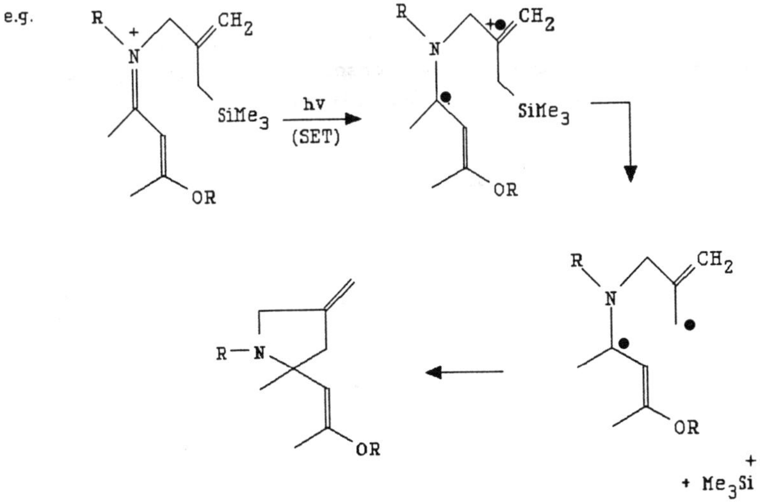

e.g. [reaction scheme showing iminium/enamine with pendant allylsilane (SiMe$_3$) undergoing hv (SET) to radical cation, then loss of Me$_3$Si$^+$ to give a diradical which cyclises to a pyrrolidine product]

+ Me$_3$Si$^+$

Interestingly, Mariano et al. recognised at an early stage that the radical cations of allylic silanes had the hole localised on the olefinic double bond and that loss of silicon required some nucleophilic assistance [13]. A similar situation apparently pertains for benzylic silicon radical cations [14].

We now report upon the free radical initiated polymerisation of acrylates in which allylic and benzylic stannanes and silanes have been used as photoinitiators in the absence and presence of electron acceptors.

Table 1 shows the number of passes under a mercury lamp that are required to cure films of an acrylate to the tack free state, using allylic and benzylic stannanes as initiators. Tables 2 & 3 show the data, obtained in a similar fashion for allylic and benzylic silanes. In the absence of sensitiser, using the full spectral output of the lamp and reducing oxygen inhibition to a minimum (the coating was covered with a quartz plate), curing was found to be efficient. When a glass plate is used to cover the film, the stannanes which lead to the most efficient cure are those which show some absorption at >300 nm, i.e. benzylic and cinnamyl stannanes. From Tables 2 & 3 it can be seen that similar results were obtained for the silanes. We attribute initiation as being caused via the silyl and stannyl radicals generated by direct excitation of the initiators.

When uncovered films were used, oxygen inhibition was so efficient that little cure was observed and this suggests that the initiator system is not particularly efficient.

$$ArCH_2MR_3 \xrightarrow{h\nu} Ar\overset{\bullet}{C}H_2 + R_3\overset{\bullet}{M}$$

$$R^1\diagup\!\!\!\diagdown\!\!\!\diagup MR_3 \xrightarrow{h\nu} R^1\diagup\!\!\!\diagdown\!\!\!\diagup^{\bullet} + R_3\overset{\bullet}{M}$$

M = Si or Sn   R = Alkyl R^1 = Ph or Allyl

Some New Developments in Radiation Curing 19

TABLE 1

Number of passes underneath a UV lamp required to cure trimethylolpropane triacrylate sensitised by allyl and benzylstannanes[a] under quartz (Q) and glass (G) in the presence and absence of sensitisers[b].

STANNANE	NO SENSITISER		A		B		C	
	Q	G	Q	G	Q	G	Q	G
 ~~SnBu₃	2	14	3	6	1	2		
 ~~SnBu₃ (methallyl)	2	13					3	7
 Ph~~SnBu₃	1	5	1	2				
PhCH₂SnBu₃	1	5	1	4			1	3
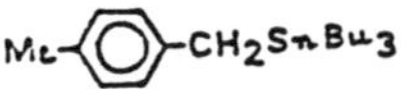 Me-C₆H₄-CH₂SnBu₃	1	3	1	3				

[a] Used at 5% w/w   [b] Used at 1% w/w

Sensitiser

A: 2,4,6-triphenylthiopyrylium BF₄⁻

B: bis(4-methoxyphenyl)-4-(4-tert-butylphenyl)thiopyrylium BF₄⁻

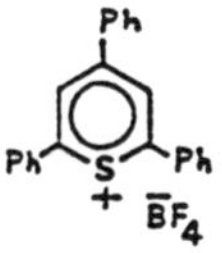

C: 9,10-dicyanoanthracene

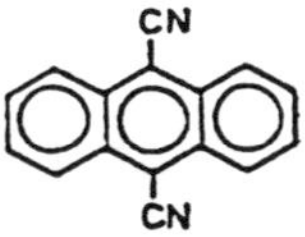

TABLE 2

Number of passes underneath a UV lamp required to cure trimethylolpropane triacrylate sensitised allyl and benzylsilanes[a] under quartz in the presence and absence of sensitisers[b].

SILANE	NO SENSITISER	A	B
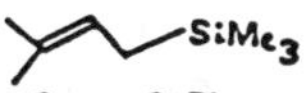 ~~SiMe₃ (methallyl)	3		3
 ~~SiPh₃	6		
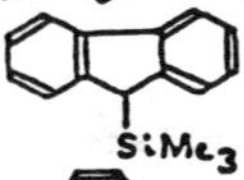 fluorenyl-SiMe₃	3		3
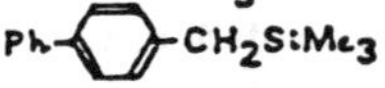 Ph-C₆H₄-CH₂SiMe₃	2	1	1
PhCH₂SiMe₃	4	1	2
Ph₂CHSiMe₃	7	3	2

[a] Used at 5% w/w   [b] Used at 1% w/w

Sensitiser

A: bis(4-methoxyphenyl)-4-(4-tert-butylphenyl)thiopyrylium BF₄⁻

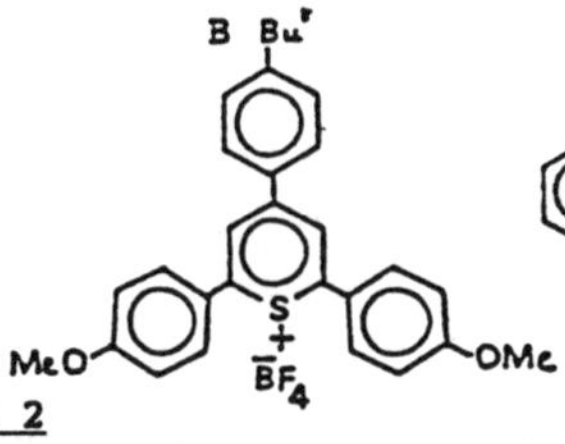

B: 9,10-dicyanoanthracene

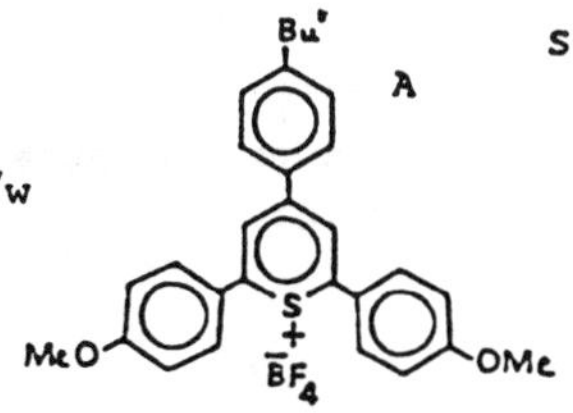

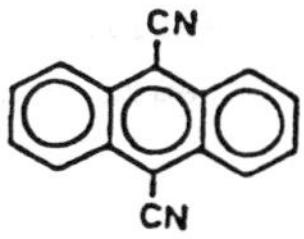

The beneficial effect of incorporating a sensitiser which acts as an electron acceptor into the system can be seen from Table 1 (results using films covered with glass) and Table 3. Although in some cases use of a sensitiser for curing films under quartz (Tables 1 & 2) led to fewer passes being required for cure, this was not always the case. Analysis of this situation is difficult since it is not easy to determine the extent to which the organometallic is being excited compared with the initiator.

Why should electron acceptors act as sensitisers? According to precedent [9-14] we should expect the following reactions to occur.

$$\text{Sens}^* + R^1MR_3 \longrightarrow \left[\text{Sens}^{\bullet-} \quad R^1MR_3^{+\bullet} \right]$$

$$\text{Sens}^{\bullet-} + R_3M^+ + R^{\bullet 1} \longleftarrow \left[\text{Sens}^{\bullet-} \quad R_3M^{+*} \quad R^{\bullet 1} \right] \longrightarrow \text{In cage combination products}$$

R^1 = allyl or benzylic group   M = Si or Sn   R = Alkyl

If this is the whole story we have a situation where the only species capable of initiation is the inefficient allylic or benzylic radical! We therefore looked at the fragmentation of some silyl and stannyl radical cations by e.p.r. spectroscopy. The silanes and stannanes in a solid fluorotrichloromethane matrix (at 77K) were exposed to γ-radiation (1Mrad) using a ^{60}Co source. Following recording of the spectra, the temperature of the matrix was gradually increased up to the melting point of the matrix (~160K) and spectra recorded, where possible, of breakdown products [15].

E.p.r. spectra of the radical cations benzyl trimethylsilane and its 4-methyl and 4-methoxy derivatives at 77K were recorded and showed that most of the spin density resided on the aromatic ring and a small amount is located in the silicon to benzylic carbon bond. On warming the matrix no direct evidence could be obtained for the formation of benzylic radicals. This observation substantiates the earlier claim that nucleophilic attack at silicon is required to cause fragmentation [13,14].

Some New Developments in Radiation Curing

TABLE 3

Number of passes underneath a UV lamp required to cure trimethylolpropane triacrylate sensitised allyl and benzylsilanes[a] under glass in the presence and absence of sensitisers[b].

SILANE	NO SENSITISER	A	B
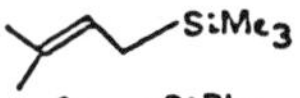 $\diagdown\!\!\diagup\!\!\diagdown$ SiMe$_3$	13		7
 $\diagdown\!\!\diagup\!\!\diagdown$ SiPh$_3$	6		
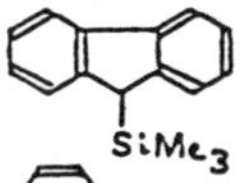 (fluorenyl)SiMe$_3$	7		6
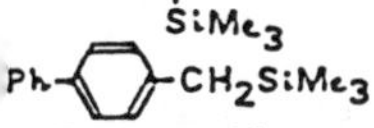 Ph–C$_6$H$_4$–CH$_2$SiMe$_3$	11	2	3
PhCH$_2$SiMe$_3$	28	3	10
Ph$_2$CHSiMe$_3$	*	13	3

[a] Used at 5% w/w   [b] Used at 1% w/w   * No cure observed

Sensitiser

A: (4-Bu^t-phenyl, 4-MeO-phenyl, 4-MeO-phenyl substituted thiopyrylium BF$_4^-$)

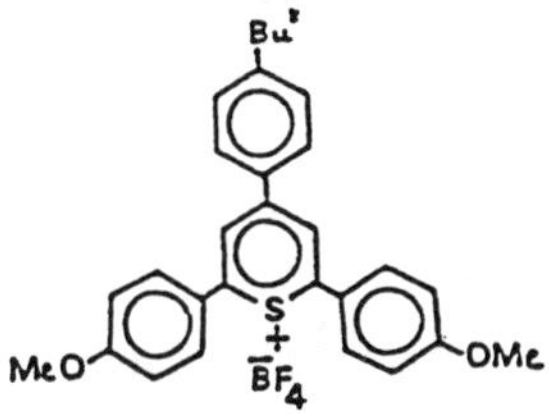

B: 9,10-dicyanoanthracene

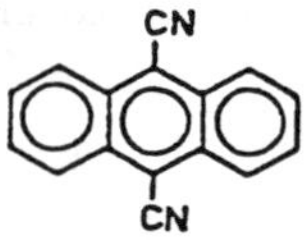

TABLE 4

Number of passes underneath a UV lamp required to cure an Epoxide[a] by iron-arene complexes, showing the beneficial effect of oxygen.

INITIATOR (1% w/w)	QUARTZ[b] COVERSLIP	GLASS[b] COVERSLIP	CURE IN AIR	GLASS FILTER
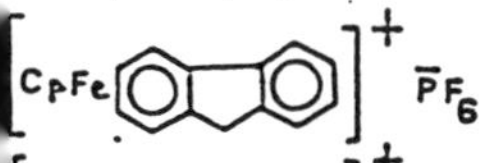 [CpFe(fluorene)]$^+$ PF$_6^-$	19	19	4	4
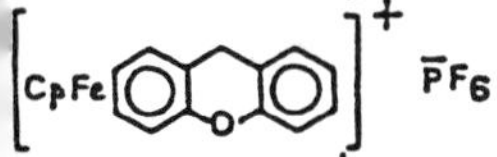 [CpFe(dibenzofuran)]$^+$ PF$_6^-$	25	25	5	6
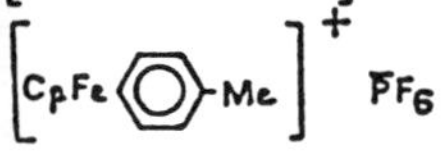 [CpFe(toluene)]$^+$ PF$_6^-$	11	11	2	4

a = (3,4-epoxycyclohexylmethyl 3,4-epoxycyclohexanecarboxylate)

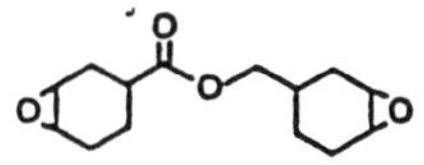

b = Under these conditions it was noticed that cure initially occurred around the edges.

Benzyl and 4-methoxybenzyl tri-n-butylstannyl radical cations were observed in solid fluorotrichloromethane and on warming the matrix, fragmentation to give the appropriate benzyl radical was observed. This fragmentation process was found to be thermally reversible for the 4-methoxy derivative.

$$[MeO-C_6H_4-CH_2SnBu_3]^{+\bullet} \underset{Cool}{\overset{Warm}{\rightleftarrows}} MeO-C_6H_4-\overset{\bullet}{C}H_2 + Bu_3Sn^+$$

Allyl and cinnamyl tri-n-butyl stannanes yielded both radical cations and the carbon centred radicals respectivity. The radical cations of these compounds fragment easier than the benzylstannyl radical cations and this probably reflects the greater stability of the carbon centred radicals produced by fragmentation. There was also evidence (observation of the fluorodichloromethyl radical) that fragmentation of the allyl tri-n-butylstannyl radical can occur in two different ways, i.e.

$$[CH_2=CH-CH_2-SnBu_3]^{+\bullet} \longrightarrow \begin{cases} CH_2=CH\overset{\bullet}{C}H_2 + Bu_3Sn^+ & \text{Major route} \\ CH_2=CHCH_2^+ + Bu_3\overset{\bullet}{Sn} & \text{Minor route} \end{cases}$$

$$Bu_3\overset{\bullet}{Sn} + FCCl_3 \longrightarrow Bu_3SnCl + \overset{\bullet}{C}FCl_2$$

This is an important observation in the context of the u.v. curing results we have described. It would seem that radical cations of the type $[R^1 MR_3]^{+\bullet}$ have two possible fragmentation routes.

$$[R^1MR_3]^{+\bullet} \longrightarrow \begin{cases} \overset{\bullet}{R}^1 + R_3M^+ & \text{Major} \\ R^{1+} + R_3\overset{\bullet}{M} & \text{Minor} \end{cases}$$

The major pathway is of no benefit to the curing process whereas the minor pathway is of significance. From our e.p.r. studies we cannot tell if the thermal energy which is available to radical cations at room temperature is able to increase the favourability of cleavage to give metal centred radicals.

Our results on the photocleavage of carbon-silicon and carbon-tin bonds suggest that these reactions could be employed to develop deep u.v. resists.

(ii) Iron arene complexes as photoinitiators.

Iron arene complexes are finding increasing use as photoinitiators for cationic curing [16,17]. For the curing of epoxides it is recommended that a "thermal bump" be applied so that tack-free coatings can be obtained within a reasonable cure time. It was also found that oxidation of the iron arenes by cumene hydroperoxide led to a dramatic increase in cure rate. We have now found that the cure of an epoxide using various iron arene complexes occurs much faster in the presence of air than in its absence. The number of passes under the u.v. lamp to give tack-free coatings are shown in Table 4. We ascribe the beneficial effect of air being due to oxygen oxidising the iron species either in the ground or excited state (it possesses a relatively long lived triplet state [18]). The iron (III) state produced in this way is supposed to promote the reaction by being a better ligator which assumes that reaction occurs by loss of the arene and it being replaced by three epoxide groups. Undoubtedly there is some evidence which favours this mechanism [17] but recent work has shown that iron arene complexes can decompose, without interacting with the epoxides, to give hydrogen fluoride [19].

It has been claimed that iron-arene complexes will also initiate free radical polymerisation reactions [20]. We have confirmed this finding (Table 5). In this case, the presence of oxygen appears to have a deleterious effect, presumably by scavenging initiator and growing polymer radicals. Use of the borate salts of iron arene complexes [21] also leads to cure (Table 6) and we speculate that with these compounds, the main initiating radical is derived from the borate anion.

TABLE 5

Number of passes[a] underneath a U.V. lamp required to cure a film (6μm) of tripropyleneglycol diacrylate initiated by iron arene hexafluorophosphates.

NUMBER OF PASSES

INITIATOR (1% w/w)	QUARTZ COVERSLIP	GLASS COVERSLIP	CURE IN AIR	GLASS FILTER
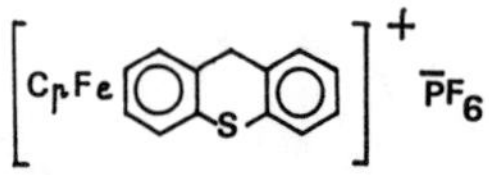 [CpFe(thianthrene)]$^+$ $\bar{P}F_6$	12	20	28	>30
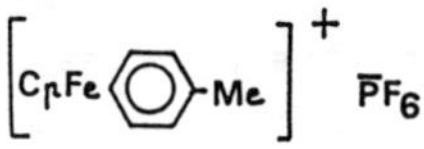 [CpFe(C$_6$H$_5$Me)]$^+$ $\bar{P}F_6$	10	19	>30	>30
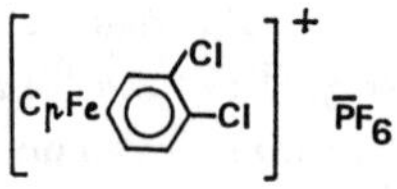 [CpFe(C$_6$H$_4$Cl$_2$)]$^+$ $\bar{P}F_6$	6	25	>30	>30

a Belt Speed 5ft min^{-1}

TABLE 6

Number of passes[a] under a UV lamp required to cure a film (6μm) of trimethylolpropane triacrylate initiated by iron arene butyl triphenylborates.

NUMBER OF PASSES [b]

INITIATOR (1% w/w)	QUARTZ COVERSLIP	GLASS COVERSLIP	CURE IN AIR	GLASS FILTER
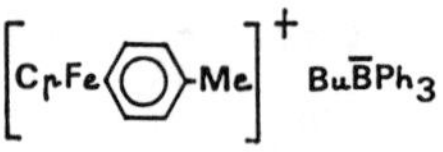 [CpFe(C$_6$H$_5$Me)]$^+$ Bu$\bar{B}$Ph$_3$	4	8	30+(4)	30+(12)
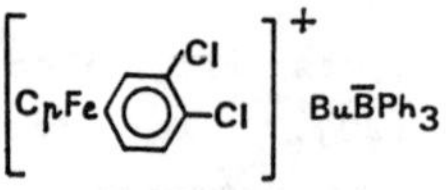 [CpFe(C$_6$H$_4$Cl$_2$)]$^+$ Bu$\bar{B}$Ph$_3$	3	8	30+(4)	30+(11)
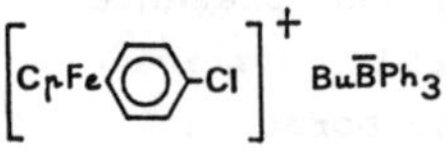 [CpFe(C$_6$H$_5$Cl)]$^+$ Bu$\bar{B}$Ph$_3$	4	9	20(4)	30+(12)

a Belt Speed 5ft min^{-1}

b In the presence of 10% N-methyldiethanolamine. Amine + TMPTA without initiator requires >50 passes to cure.

$$[\text{CpFeArH}]^+ + \text{BuBPh}_3^- \xrightarrow[\text{SET}]{h\nu} \text{CpFeArH} + \text{Bu}^\bullet + \text{Ph}_3\text{B}$$

$$\text{Bu}^\bullet + \text{Acrylate} \longrightarrow \text{Polymer}$$

Our results suggest that the iron-arene complexes are not particularly efficient radical initiators. However, if a dual cure system based on an epoxide-acrylate mixture, is to be used many of the cationic photoinitiators including iron-arene complexes may well lead to the cure of both components.

(iii)   Quinoxalines as photoinitiators.

The quinoxaline nucleus is to be found in a number of dyes and closely related structures are found in riboflavin and dyes. Many of these materials show poor light fastness due to their ability to form excited states which abstract hydrogen from suitable C-H bonds. Quinoxalines like quinolines will induce free radical polymerisation via hydrogen abstraction from alkanes [22].

Sometime ago we noticed that simple quinoxalines on irradiation in degassed solutions containing hydrogen donors such as triethylamine underwent reduction. On admission of air the quinoxaline was regenerated. A simplified mechanism which explains such phenomena is shown in Scheme 1.

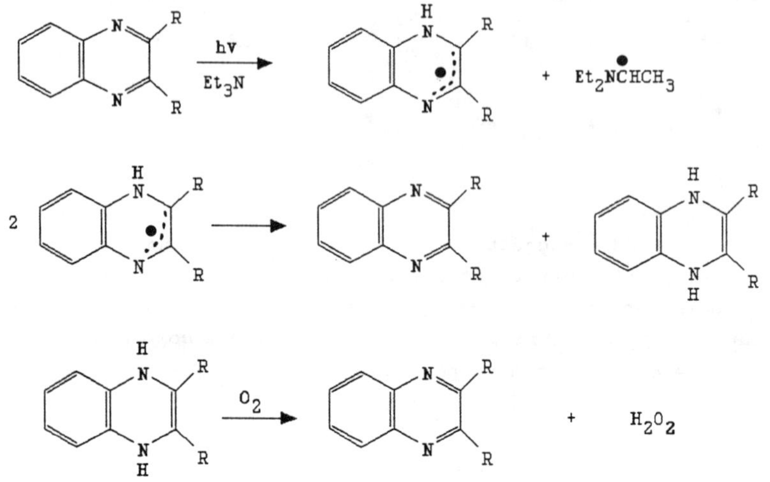

SCHEME 1

If oxygen is present in solution during irradiation it is possible that it will react with the semi reduced quinoxaline.

It will be noted that hydrogen abstraction (probably via electron transfer followed by proton transfer) from the tertiary amine leads to an α-amino alkyl radical. Such radicals are known to be good initiators for acrylate polymerisation [23].

Various quinoxalines, in association with N-methyl-diethanolamine as synergist, were used to initiate the polymerisation of acrylates and some of the results are shown in Table 7. As can be seen from Table 7, the quinoxalines when used with a good amine synergist such as N-methyldiethanolamine are efficient Type II initiators. The simplicity of synthesis of these materials and the ease with which their long wavelength band can be moved to the red by an appropriate substitution means that they can be readily tailored to operate in the visible region.

That oxidation of the reduced quinoxalines by molecular oxygen plays a part in the curing process has been demonstrated by using the technique of Real Time Infrared Spectroscopy (RTIR). It was found that the cure of the acrylate is faster when oxygen is present in the reaction mixture than when it is absent.

TABLE 7

Number of passes[a] under a UV lamp[b] required to produce a tack free film of trimethylolpropane triacrylate using a mixture of a quinoxaline and N-methyldiethanolamine as the initiator system.

QUINOXALINE (M)	NUMBER OF PASSES
(structure 1)	6[c] 1[d]
(structure 2, dimethyl)	8[c]
(structure 3, morpholino)	1[d]
(structure 4, diphenyl)	7[c] 2[d]
(structure 5)	7[c]
(structure 6)	8[c]

a Belt Speed = 5ft min⁻¹
b Full output of medium pressure mercury lamp (rated at 100W. per inch used).
c 0.1% initiator, 3% amine.
d 1% initiator, 10% amine, belt speed 100ft/min.

TABLE 8

Bulk polymerisation experiments using various initiators and amine synergists and trimethylolpropane triacrylate as monomer.

INITIATOR	SYNERGIST	REACTION TIME (MIN)	% POLYMER
ITX[a]	IPM[b]	6.28	7.7
ITX	NMDEA[c]	6.28	74
Irgacure 907	NIL	6.36	80
Irgacure 907	IPM	6.25	80
Irgacure 907	NMDEA	6.25	90
Benzophenone	IPM	6.28	Zero
Benzophenone	NIL	6.28	Zero
Benzophenone	NMDEA	6.28	70

a Isopropylthioxanthone (0.02%)
b Isopropylmorpholine (1.1%)
c N-Methyldiethanolamine (1.3%)

In accordance with previous work [23] we find that many of the quinoxalines will initiate the polymerisation of acrylates in the absence of amines, but oxygen inhibition of cure is very marked.

(iv) The photoinitiator 1-(4-methylthiophenyl)-2-methyl-2-(N-morpholino) propan-1-one. (Irgacure^T 907)

This Type I photoinitiator is both a useful and interesting material. It has been shown by CIDNP [24] radical trapping experiments [25] and CIDEP experiments [26] that it fragments in the following fashion.

$$CH_3S-C_6H_4-C(=O)-C(CH_3)_2-N(morpholino) \xrightarrow{h\nu} CH_3S-C_6H_4-C(=O)\bullet + \bullet C(CH_3)_2-N(morpholino)$$

The triplet state of the ketone (from which reaction occurs as demonstrated by CIDNP experiments) has an estimated lifetime of ≤10 n secs [27]. If used in curing experiments at the 3% level (this equates to a concentration of ~10^{-1} M) it is possible that some bimolecular reactions may ensue between a molecule in its triplet state and one in its ground state. Since the initiator is an amine this raised the question as to whether the initiator can act as its own amine synergist.

i.e.

$$Ar-C(=O)-C(CH_3)_2-N(morpholino) \xrightarrow{h\nu} ArC(=O)\bullet \,\, \bullet C(CH_3)_2-N(morpholino) \,\, / \,\, ArC(=O)-C(CH_3)_2-\overset{+\bullet}{N}(morpholino)$$

Ar = CH$_3$S–C$_6$H$_4$–   Quenching

$$Ar-C(OH)(CH_3)-C(CH_3)_2-N(morpholino) \,\, + \,\, ArC(=O)-C(CH_3)_2\bullet \,\, N(morpholino)$$

Initiates polymerisation

Such a scenario has in fact been described and the claim made that the morpholino group acts as a hydrogen donor [26].

We have studied the synergistic properties of N-isopropylmorpholine (IPM) to test this claim. In bulk polymerisation experiments (Table 8) it was found that little polymerisation of trimethylolpropane triacrylate (TMPTA) occurred when IPM was used as synergist with the efficient Type II initiator isopropylthioxanthone. When N-methyldiethanolamine was used efficient polymerisation was observed. Irgacure 907, being a Type I photoinitiator, initiated the polymerisation of TMPTA efficiently in the absence of amine. Addition of NMDEA had a marginal beneficial effect (probably due to oxygen scavenging) where IPM failed to produce any effect.

That IPM is a grossly inefficient synergist was proved by experiment using Type II photoinitiators (thioxanthones and benzophenone) in conjunction with IPM for the curing of thin films.

Using another technique photo-differential scanning calorimetry (photo DSC) it was shown that high concentrations of IPM can quench the Type I cleavage of Irgacure 907.

We conclude that IPM and the morpholino residue in Irgacure 907 act as physical quenchers for the triplet states of the ketone. There is an indication that the rate constant for this process may be less than diffusion controlled which can be accounted for in terms of the relatively high ionisation potential of the amine and the bulky substituent attached to the nitrogen.

ACKNOWLEDGEMENTS. Financial support from the following organisations is gratefully acknowledged: SERC for a Cooperative and CASE award with Cookson Group plc, Cookson Group plc, Horsell Graphics and Yildiz University, Chemistry Department.

We are also grateful to Dr J.D. Coyle of Cookson Group plc for most stimulating discussions.

REFERENCES

1. M.S. Salim in 'Radiation Curing of Polymers II Ed. D.R. Randell, Royal Society of Chemistry 1991, 3.
2. EEC Directive 90/128/EEC relating to plastic materials and articles intended to come into contact with foodstuffs.
3. For a discussion of e.b. curing see R.S. Davidson in 'Radiation Curing in Polymer Science and Technology' Ed. J.P. Fouassier and J.F. Rabek. Elsevier Applied Science UK 1993. To be published.
4. For a recent review of the development of new photoinitiators over the last five years see R.S. Davidson, J. Photochem. Photobiol A. Chem Edn. In the press.
5. J.M.J. Frechet, S. Matuszczak, H.D.H. Stover, C.G. Willson and B. Reck, ACS Symposium Series 1984 *412* 74.
6. R.J. Batten, R.S. Davidson, R.J. Ellis and S.A. Wilkinson Polymer 1992 *33* 3037.
7. R.S. Davidson, R.J. Ellis and S.A. Wilkinson Polymer 1992 *33* 1836.
8a. C. Chatigilialoglu, K.U. Ingold and J.C. Scaiano, J. Amer. Chem. Soc., 1983, *105*, 3292.
8b. W.B. Motherwell, A.M.K. Pennell and F. Ujjainwalla, Chem. Comm., 1992, 1067; G. Stork and R. Mook, J. Amer. Chem. Soc., 1987 *109* 2828.
9. D.F. Eaton, J. Amer. Chem. Soc. 1981, *103* 7235.
10. S. Kyushin, Y. Masuda, K. Matsushita, Y. Nakadaira and M. Ohashi, Tetrahedron Letters 1990, *31* 6395.
 For similar reactions see:
 K. Mizuno, K. Terasaka, M. Yashueda and Y. Otsuji, Chem. Letters 1988, 145.
 S. Kyushin, Y. Nakadaira and M. Ohashi. Chem. Letters 1990 2191.
 A. Sulpizio, A. Albini, N. d'Alessandro, E. Fasani and S. Pietra, J. Amer. Chem. Soc. 1989 *111* 5773.
 N. d'Alessandro, E. Fasani, M. Mella and A. Albini, J. Chem. Soc. Perkin II 1991, 1977.
11. A. Takuwa, Y. Nishigaichi, K. Yamashita and H. Iwamoto Chem. Letters 1990, 639, 1761.
 A. Takuwa, Y. Nishigaichi, T. Yamaoko and K. Iihama J.C.S. Chem. Comm., 1991, 1359.
12. I-S Cho. C-L Tu and P.S. Mariano, J. Amer. Chem. Soc., 1990 *112* 3594.
13. K. Ohga and P.S. Mariano, J. Amer. Chem. Soc. 1982 *104* 617.

14. J.P. Dinnocenzo, S. Farid, J.L. Goodman, I.R. Gould, W.P. Todd and S.L. Mattes, J. Amer. Chem. Soc. 1989 $\underline{111}$ 8973.
    S.R. Sirimanne, Z. Li, D.R. Vander Veer and L.M. Talbot, J. Amer. Chem. Soc. 1991, $\underline{113}$ 1766.
15. E. Butcher, C.J. Rhodes, M. Standing, R.S. Davidson and R. Bowser, J.C.S. Perkin II, 1992, 1469.
16. A. Roloff, K. Meier and M. Riediker, Pure and Applied Chemistry, 1986, $\underline{58}$ 1267.
17. K. Meier and H. Zweifel, J. Radiation Curing 1986, (October) 26.
18. D.R. Chrisope, K.M. Park and G.B. Schuster, J. Amer. Chem. Soc., 1989 $\underline{111}$ 6195.
19. K.W. Allen, E.S. Cockburn, R.S. Davidson, K.S. Tranter and H.S. Zhang, Pure and Applied Chem, 1992 $\underline{64}$ 1225.
20. K. Yamashita and S. Imahashi, Jpn. Kokai Tokkyo Koho JP02305806 C Chem. Abs. 1142093302.
21. D.R. Chrisope and G.B. Schuster Organometallics, 1989 $\underline{8}$ 2737.
22. D. Braun Angewandte Makromol. Chem. 1990 $\underline{183}$ 17.
23. For a discussion on amines as synergists see R.S. Davidson in 'Radiation Curing in Polymer Science and Technology', ed. J.P. Fouassier and J.F. Rabek, Elsevier Applied Science UK 1993. To be published.
24. V. Desobry, K. Dietliker, R. Husler, W. Rutsh and H. Loeliger, Polymers Paint Colour Int, 1988 $\underline{178}$ 913.
25. K. Meier, M. Rembold, W. Rutsh and F. Sitek in 'Radiation Curing of Polymers' ed. D.R. Randell, Royal Society of Chemistry, 1987, 196.
26. J.P. Fouassier, D.J. Lougnot, A. Paverne and F. Wieder, Chemical Phys. Letters 1987, $\underline{135}$ 30.
27. J.P. Fouassier and D. Burr, Eur. Polym. J. 1991 $\underline{27}$ 657.

Photo-induced Polymerization – Kinetic Study of Ultrafast Processes Induced by UV Radiation or Lasers

C. Decker

LABORATOIRE DE PHOTOCHIMIE DES POLYMÈRES (CNRS), ÉCOLE NATIONALE SUPÉRIEURE DE CHIMIE, 68200 MULHOUSE, FRANCE

1. INTRODUCTION.

The light-induced polymerization of multifunctional monomers is one of the most efficient methods to produce quasi-instantly highly crosslinked polymers. With the widely used acrylic resins, the chain reaction develops extensively under intense illumination, leading within milliseconds to a totally insoluble material. The UV-curing technology has found a large variety of applications, mainly in the coating industry[1], in the graphic arts[2] and in microelectronics[3], due to its distinct advantages: great cure speed, solvent-free formulations, room temperature operations, selective polymerization in the illuminated areas and tailor-made properties of the cured polymer.

In recent years, a number of new products have been developed in order to further improve the performance of UV-curable systems, with respect to both the resin reactivity and the properties of the final product. Substantial progress has thus been achieved by using some highly efficient photoinitiators[4] and very reactive monomers and oligomers[5]. In the continuing search for ever larger cure rates, a great step forward has been taken by replacing the traditional mercury or xenon lamp by powerful lasers, which appear today as the ultimate light sources to provide quasi-instantaneous polymerization[6-7]. Exposure times as short as a few nanoseconds proved to be sufficient to transform the liquid resin into a solid, insoluble polymer, and to do it selectively in well-defined areas of micronic dimension. Owing to its unique advantages, laser-induced polymerization has found its main end-uses in areas where cure speed and spatial selectivity are of major concern, like in direct imaging[8], stereolithography[9] and holography[10].

One of the major problems encountered when studying such ultrafast reactions is to find a reliable method that would allow one to monitor continuously and quantitatively polymerization processes which occur in a fraction of a second, and thus to determine the important kinetic parameters. In this paper, we describe the performance of a recently developed method[11], based on real-time infrared (RTIR) spectroscopy, which proved particularly well suited to monitor ultrafast UV-or laser-induced polymerizations, and evaluate the reactivity of some novel photoinitiators and acrylic monomers. When used in UV-curable formulations, these compounds were found to be highly efficient for achieving fast and complete cure, while in addition they improved substantially the mechanical properties of the final product.

2. EXPERIMENTAL.

Materials.

The photopolymerizable resins used in this work contained three main components :
- a photoinitiator that cleaves readily upon UV or laser exposure to generate free radicals. For most experiments, α,α'-dimethoxy-phenylacetophenone (Irgacure 651 from Ciba-Geigy) was selected because of its high efficiency in producing free radicals upon photolysis (mostly benzoyl and methyl radicals), and its wide-spread use in UV-curing applications.
- an acrylate end-capped oligomer consisting of an aliphatic polyurethane (Actilane 20 from SNPE), a polyester (Ebecryl 80 from UCB), or a derivative from the glycidyl ether from bis-phenol A (Actilane 72 from SNPE).
- an acrylic monomer which serves as a reactive diluent, such as ethyldiethyleneglycol acrylate (EDGA from SNPE), hexanediol diacrylate (HDDA from UCB), or trimethylolpropane triacrylate (TMPTA from UCB). Some recently developed mono-acrylates containing a carbamate, oxazolidone or cyclic carbonate function in the monomer unit have also been evaluated for their performance in UV-curable resins. The formulae and trade names of these compounds from SNPE are given in Fig.1.

Typical formulations contained up to 5 wt.% of photoinitiator and equal parts of the acrylic monomer and functionalized oligomer. The resin was applied on a KBr crystal as a uniform layer of controlled thickness, usually 20 μm, with a calibrated wire-wound applicator.

UV-Exposure.

Two types of light sources were employed to induce the polymerization reaction :
- a medium pressure mercury lamp (HOYA-HLS 210 U), whose emission spectrum extends from 248 to 578 nm. The radiation was focused and directed towards the sample by means of an optical fiber.
- a krypton ion laser (spectra Physics Model 2020), tuned to its UV emission line at 337.4 nm and operated in the continuous wave mode. The power output of the laser could be varied continuously, up to a maximum value of 300 mW.

The incident light-intensity at the sample position was measured by a radiometer (International Light), and could be adjusted to any value in the range of 10 to 1400 mW cm^{-2}. The duration and exact start of the exposure were determined precisely by means of an electronic shutter and a photocell connected to a transient memory recorder.

The photopolymerization experiments were carried at room temperature, in the presence of air. In order to prevent atmospheric O_2 from diffusing into the sample during the irradiation, some experiments were carried out with laminates, the resin

Fig. 1 : Formulae of new acrylate monomers

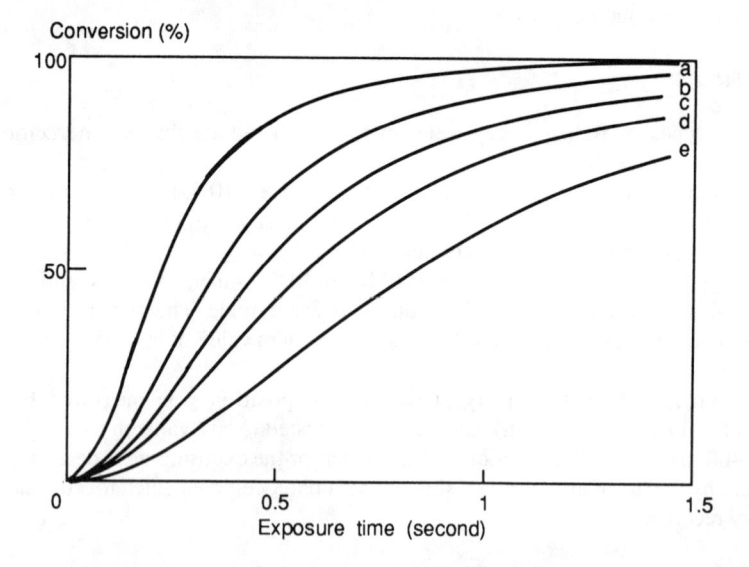

Fig. 2 : Polymerization profiles recorded by RTIR spectroscopy for a polyester-acrylate exposed as a laminate to UV radiation in the presence of different types of photoinitiators. [Photoinitiator] : 1% ; light intensity : 60 mW cm^{-2} ; a : Irgacure 369 ; b : Irgacure 907 ; c : isopropyl thioxanthone ; d : Darocure 1173 ; e : benzophenone.

coated KBr crystal being covered with a polyethylene film.

Analysis by Real-Time Infrared (RTIR) Spectroscopy.

The method used to study in real time the kinetics of ultrafast light-induced polymerizations has already been described for both acrylic[12] and epoxy monomers[13]. The sample was exposed simultaneously to the UV beam, which induces the polymerization, and to the analyzing IR beam, which monitors continuously the drop in the IR absorbance of the reactive group. The spectrophotometer was operated in the absorbance mode and the detection wavelength set at 812 cm^{-1} where acrylic monomers exhibit a sharp absorption band (CH$_2$ = CH twisting). The decrease of the sample absorbance upon UV exposure was monitored in real time on a transient memory recorder. Since the IR absorbance is proportional to the monomer concentration, the kinetic curves thus recorded can be directly translated into conversion vs. time profiles. Due to its short response time (40 ms), this technique permits a correct kinetic analysis of polymerization reactions proceeding in a timescale as short as 100 ms.

A similar method, based on real time ultraviolet (RTUV) spectroscopy, was used to monitor the disappearance of the photoinitiator (PI) upon irradiation of UV-curable resins[14]. From the PI loss profiles that were directly recorded, photolysis rate constants and quantum yields were evaluated, for both radical-and cationic type photoinitiators.

3. KINETIC STUDY OF LIGHT-INDUCED POLYMERIZATION.

Evaluation of kinetic parameters

The analytical methods currently used to study the kinetics of photopolymerization reactions can be ranged in two main classes, depending on their mode of operation :
- those based on discrete measurements of the chemical modifications occurring after a short exposure to a high intensity UV source (infrared spectroscopy, photoacoustic spectroscopy, NMR spectroscopy, gravimetry).
- those based on the continuous monitoring in real time of some physical modifications induced by UV-curing, like the heat evolved (photocalorimetry, IR radiometry), the shrinkage (dilatometry), or changes in the refractive index (laser-interferometry), or in the microviscosity (fluorimetry).

The former methods provide quantitative data on the rate of polymerization and on the actual amount of residual monomer in the final product, but they require a great number of measurements to be made, and are therefore ill-suited for mass production analysis. In this respect, real-time techniques present the distinct advantage of allowing a continuous monitoring of the curing process, and thus a fast performance analysis of various photosensitive resins. They suffer however two important limitations which restrict their potential use for analyzing ultrafast polymerization processes. Some of them, like dilatometry or the widely used differential photocalorimetry (DPC), have an inherent response time which is much too long to allow a reliable study of curing reactions that develop extensively in less than one second. The others, like interferometry or fluorimetry, have a fast response, but they have the major disadvantage of providing no quantitative information on the polymerization rate or cure extent, and are therefore limited to qualitative investigations.

In order to be able to follow both continuously and quantitatively ultrafast polymerization reactions, we have transformed the infrared spectroscopy method into a real-time technique. By exposing the sample simultaneously to the UV beam and to the IR beam, conversion vs. time curves have been directly recorded, for polymerizations that proceed extensively within a few tenths of a second under exposure to UV radiation or laser beams[15]. Fig. 2 shows some typical polymerization profiles recorded by RTIR spectroscopy upon UV exposure in the presence of air of a polyester - acrylate resin containing different types of photoinitiators. It can be seen that, with the most efficient initiator, close to 100 % conversion was reached within 1 s of irradiation.

The S shape curve observed results from two main factors :
 - in the early stages, the oxygen dissolved in the sample has a strong inhibiting effect, due to its reaction with both the initiator and the polymer radicals,
 - in the later stages, the depletion of the monomer, together with the gel effect and its segmental mobility restrictions, leads to a slowing down of the reaction.

One of the distinct advantages of RTIR spectroscopy is that the important kinetic parameters can be directly evaluated in a single experiment, like the rate of the polymerization (R_p) and its quantum yield ($\emptyset_p$), the photosensitivity and the precise amount of residual monomer in the final product. The influence of the formulation constituents on these kinetic parameters was thus determined for different types of UV-curable resins, as well as the effect of some physical factors like the light-intensity, the radiation wavelength, the film thickness or the O_2 concentration[12-16]. Fig. 3 shows for example the RTIR curves recorded for a polyurethane-acrylate resin exposed to UV radiation at various intensities. A most remarkable feature is that the same polymerization profile was recorded when the light-intensity was increased above 100 mW cm^{-2}, up to 1.4 W cm^{-2}. This result clearly indicates that the actual polymerization rate reaches a maximum value (about 17 mol L^{-1} s^{-1} for the resin considered), which cannot be exceeded by further increasing the initiation rate. This saturation effect is not due to the speed limit of the signal detection, or to a fast depletion of the photoinitiator[16]. It most probably results from the fact that, as R_p reaches such high values, the limiting factor becomes the speed at which the polymer chains are actually growing.

In this respect, it should be mentioned that a very similar RTIR curve was recorded when the polymerization was induced by pulsed laser irradiation[16]. Large amounts of initiating radicals are then generated during the very short and intense flash, so that the polymerization itself takes exclusively place in the dark, just after the exposure. The RTIR profile recorded provides thus a reliable measure of the true rate at which polymerization develops, a quantity which depends primarily on the

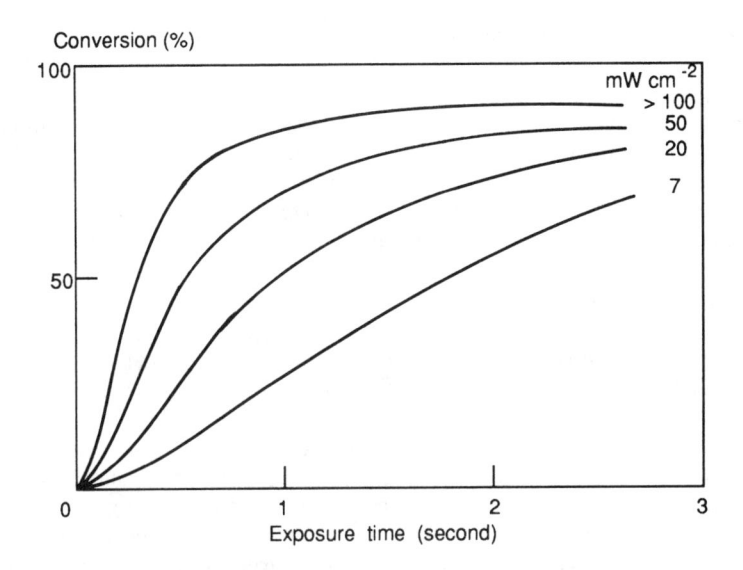

Fig. 3 : Influence of the light-intensity on the polymerization profile of a polyurethane acrylate, [Irgacure 369] = 1%

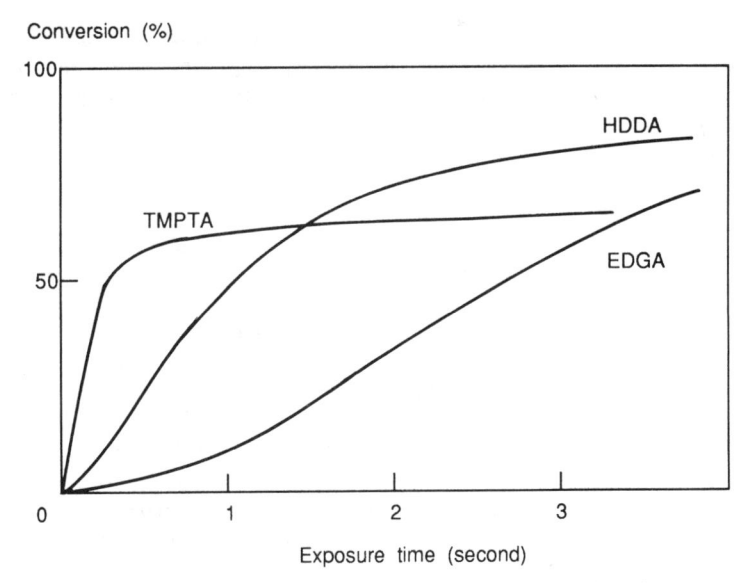

Fig. 4 : Influence of the reactive diluent on the light-induced polymerization of a polyurethane-acrylate

propagation and termination rate constants.

Compared to the other methods of kinetic analysis, RTIR spectroscopy shows the best performance, for it gives instantly access to quantitative information on photopolymerization reactions carried out under the same conditions as those found in most UV-curing applications. Real-time monitoring, sensitivity, fast response and quantitative evaluation are among its prominent advantages, which cannot be matched by any of the competitive techniques.

High Performance New Monomers

The monomer used as reactive diluent in a UV-curable resin plays a key role, for it affects both the rate of polymerization and the maximum conversion. This is clearly illustrated by Fig.4 which shows the RTIR profiles recorded for a polyurethane-acrylate resin containing a mono-, di-, or triacrylate diluent, in a 1 to 1 ratio. As the monomer functionality was increased, the polymerization developed more rapidly, due to both the higher initial concentration in acrylic groups and the greater viscosity. At the same time, the more pronounced gel effect, with its related mobility restrictions, makes the reaction stop at an earlier stage, so that there remains a larger amount (~35%) of unreacted monomer in the triacrylate-based UV-cured polymer. This highly crosslinked material proved to be much harder and scratch resistant than the monoacrylate-based coating, but also much less flexible and impact resistant.

One of the main objectives of our work was to develop some new monomers that would polymerize both rapidly and extensively, to give either soft and impact resistant elastomers, or hard but still flexible coatings. This was achieved by introducing into the structural unit of a monoacrylate one of the following functional groups : carbamate, oxazolidone, dioxolane, cyclic carbonate or chlorine (Fig.1).

When exposed to UV radiation in the presence of a photoinitiator, these monomers were found to polymerize as fast as triacrylates and as extensively as monoacrylates[17-18]. A similar behaviour was observed for their copolymerization with either polyurethane-acrylate or epoxy-acrylate oligomers[19], as shown by the high degree of conversion reached after an exposure as short as 10 ms (Fig. 5). The most reactive functional group, based on the value of the $R_p/[A]_o$ ratio, appears to be the chlorinated polyether. The polymerization rate was found to increase by a factor of 20, compared to a conventional monoacrylate like EDGA, with formation of a polymer that contains essentially no residual unsaturation. Close to 100 % degrees of conversion were also reached with the carbamate, oxazolidone, dioxolane and cyclic carbonate derivatives, at similarly high cure speeds, as shown by the $R_p/[A]_o$ values reported in Table 1. The greatly enhanced reactivity of the chlorine substituted monomer is assumed to result from both an efficient chain transfer process involving the labile hydrogens in the α and β positions, and a possible contribution of the chlorine radicals in the initiation process, as was already observed with chlorine substituted photoinitiators[20].

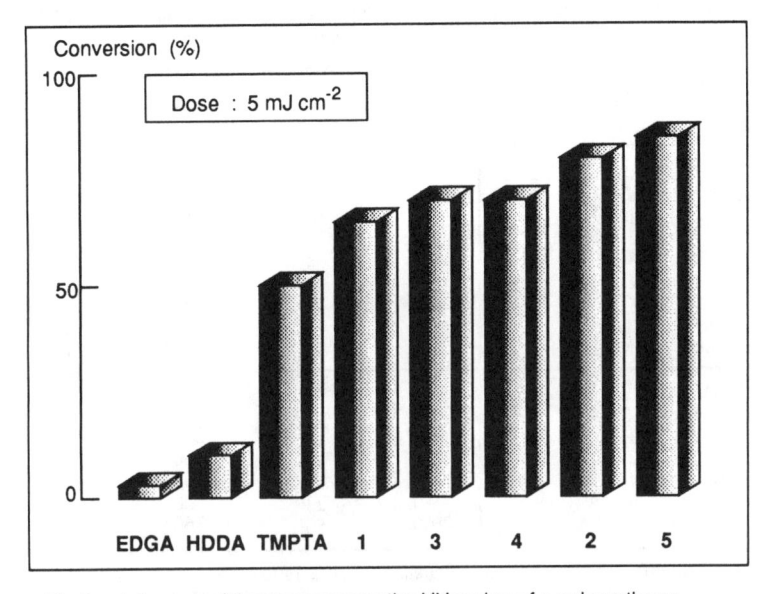

Fig.5 : Influence of the monomer on the UV curing of a polyurethane - acrylate resin. Conversion after 10 ms of UV exposure at 500 mW cm^{-2} in the presence of air

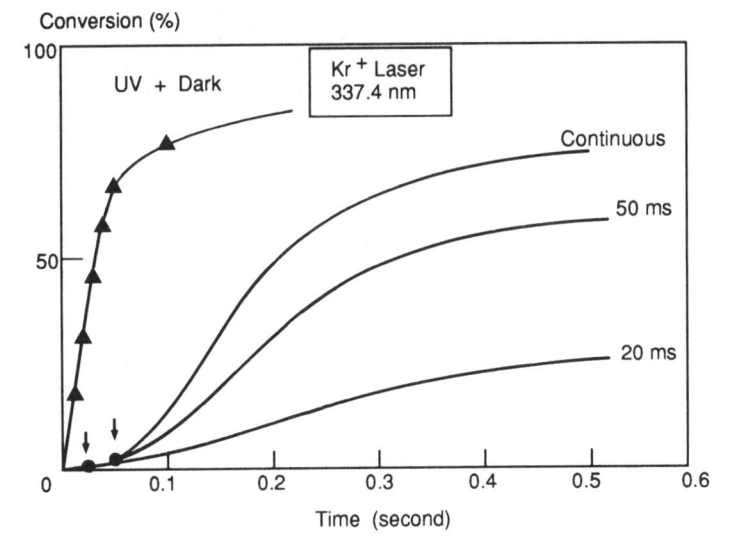

Fig. 6 : Kinetic profiles recorded by RTIR spectroscopy upon laser exposure of a polyurethane-acrylate (I = 250 mW cm^{-2})

Table 1 Performance analysis of various acrylic monomers in UV-curable resins

MONOMERS	Reactivity [a] $R_p/[A]_o$ (s-1)	Residual [b] unsaturation (%)	Persoz [b] hardness (s)	Mandrel [b] flexibility (mm)
POLYURETHANE-ACRYLATE				
Monoacrylates : EDGA	9	2	30	0
1	94	4	80	0
2	120	4	90	0
3	100	4	100	0
4	125	4	270	0
5	180	2	40	0
Diacrylate : HDDA	20	16	150	2
Triacrylate : TMPTA	110	36	270	5
POLYPHENOXY-ACRYLATE				
Monoacrylates : EDGA	8	5	60	0
1	80	9	250	1
2	110	10	260	1
4	100	19	360	1
Diacrylate : HDDA	20	23	300	12
Triacrylate : TMPTA	100	50	360	30

a) Rates of polymerization (R_p) were calculated from the ratio of the amount of acrylate double bonds polymerized to the exposure time. (light-intensity : 500 mW cm^{-2})
b) in tack-free UV-cured coatings.

When used as reactive diluents in polyurethane-acrylate resins, these new monomers give soft, highly flexible elastomers (T_g < -20° C), which have an excellent impact resistance (Table 1). The cyclic carbonate-monoacrylate **4** exhibits a distinct behavior in that it leads to hard, but still flexible coatings. When compared to the monoacrylates currently used to obtain low-modulus polymers for UV-curable adhesives or laminates, these compounds have the great advantage of polymerizing much faster, thus affording a substantial increase of the speed of the production line.

Hard and glassy polymers are usually obtained by copolymerization of multiacrylic monomers with polyphenoxy-acrylate oligomers[17-19]. On a scale which extents from 0 to 400 s for mineral glass, the Persoz hardness was found to reach values of 250 s for the carbamate and oxazolidone derivatives, and up to 360 s for the cyclic carbonate-acrylate, compared to 60 s only for conventional monoacrylates, like EDGA (Table 1). These high values are similar to those obtained by using diacrylate (HDDA) or triacrylate (TMPTA) diluents but, by contrast to such strongly crosslinked polymers which are stiff and brittle, the polyphenoxy-acrylate coatings based on monomers **1,2** and **4** were found to be highly flexible and impact resistant.

The polymerization mechanism of these new mono-acrylates is now being investigated in order to elucidate the basic reason of their outstanding reactivity. The fact that all the homopolymers were found to be strictly insoluble in the organic solvents is an indication that these monomers actually behave like difunctional crosslinking agents. Moreover, they were shown to exhibit a strong hydrogen donor character when polymerized with benzophenone as photoinitiator[21], a compound which works only in the presence of H donor molecules. These two sets of results strongly suggest that an efficient chain transfer reaction is taking place during the polymerization. This would lead to the formation of a tridimensional polymer network, and to a sharp increase of the viscosity, which may account for the faster cure.

Post-Polymerization Study

By contrast to cationic photoinitiated polymerization, which is known to further develop after the UV exposure, in radical photoinitiated polymerization the polymer chains will rapidly stop growing, once the light has been switched off, due to the terminating interactions of the propagating radicals. RTIR spectroscopy proved to be a technique particularly well suited to monitor in real time the short-lasting dark polymerization which occurs just after the irradiation.

Fig.6 shows the kinetic profiles recorded for a polyurethane-acrylate resin exposed to a krypton ion laser beam for 20 and 50 ms. It can be seen that the polymerization continues to proceed for a few seconds in the dark, due to the high concentration of initiating species generated during the short but intense illumination. This result indicates that, in such viscous media which undergo gelation, the propagation step competes favorably with the diffusion controlled termination step. It also explains the fact that, despite the high initiation rates provided by the laser irradiation, the polymerization develops with long kinetic chains *(kcl)*. From the ratio of the polymerization quantum yield to the initiation quantum yield, *kcl* values were calculated to be on the order of 10^4 mol radical^{-1} for experiments carried out in the presence of air, and up to 5×10^4 mol radical^{-1} in an inert atmosphere[22].

Fig.6 also shows the kinetic curve obtained by taking into account the polymerization which occurs both during the exposure, and afterwards in the dark. It clearly appears that most of the polymer has actually been formed just after the irradiation, especially in the early stages where post-polymerization accounts for up to 90 % of the total process [16]. This quantity depends also on the light intensity, and may even reach 100 % in the case of polymerizations induced by pulsed laser irradiation[7,23].

Some further information about the reaction kinetics can be inferred from the RTIR curve recorded during and after the UV exposure. By measuring the initial rate of the dark polymerization (R'_p), it is possible to obtain the growth profile of the radical-concentration, simply by plotting the ratio $R'_p / [M]_t = k_p [P^\bullet]$ as a function of the exposure time [16]. Under the given experimental conditions, the polymer radicals were found to reach their steady-state concentration after about 30 ms. Such RTIR curves also allow the propagation (k_p) and termination (k_t) rate constants to be determined at any stage of the reaction[24]. For the multiacrylic monomers studied, both k_p and k_t were found to decrease rapidly, once gelation occurred and the sample turned into a solid polymer film. The high reactivity of these new acrylic monomers was shown to result from both an efficient propagation step (k_p ~6.10^3 L mol^{-1} s^{-1}) and a relatively slow termination process (k_t ~2.10^4 L mol^{-1} s^{-1}).

Photo-induced polymerization in the solid state.

Although most of the photosensitive systems used in today's UV curing applications consist of liquid resins, a dry film technology may present some advantages for specific end-uses, like in photolithography and microelectronics for the fabrication of printing plates or integrated circuits. The main drawback of operating in a solid medium lies in the sharply reduced rate of polymerization, which results from severe mobility restrictions of the reactive sites. For instance, the intrinsic reactivity ($R_p/[A]_o$) was found to drop by more than two decades when the high performance Acticryl® CL 959 was photopolymerized in a poly(methylmethacrylate) matrix (in a 1 to 1 ratio), instead of in a polyurethane-acrylate resin[25].

Similar results were obtained by using as solid matrix a linear polyurethane, which became highly crosslinked within a few seconds of irradiation, to generate a tough and impact resistant material[25]. This dry film technology proved to be particularly well suited to produce safety glasses by a process faster and cheaper than the traditional thermal curing, which is carried out at elevated temperatures and under high pressure for several hours.

A most remarkable feature was observed when the UV-curing reaction was performed in a poly(vinyl chloride) matrix. The polymerization of the acrylic monomer was found to develop 100 times faster than in PMMA or polyurethane films, and almost as rapidly as in the most reactive liquid acrylic resins (Fig.7). Such a drastic rate effect strongly suggests that PVC participates in the reaction, probably as an efficient chain transfer agent through its labile secondary hydrogens. Two sets of results argue in favor of the grafting of acrylate segments, which would connect together the PVC chains : (i) the polymer becomes totally insoluble in organic solvents after UV exposure, and (ii) the glass transition temperature of the irradiated material is 50° C higher than that of PVC. Morever, the crosslinked PVC film was found to be both harder and less flexible than the original PVC.

Due to its distinct advantages, a fast dry-processing performed at ambient temperature, photopolymerization in the solid state appears as a technology well suited to modify the properties of non-functional polymers.This environment friendly process is expected to find its major applications in optoelectronics, in microlithography and in the coating industry.

Photolysis of the Initiator.

In light-induced polymerization, the photoinitiator (PI) is destroyed at a rate which depends on the light-intensity, the PI absorbance and the photolysis quantum yield. A kinetic study of the photoinitiator disappearance during UV curing will provide useful information about two important quantities : (i) the initiation rate, which depends directly from the PI loss rate and, (ii) the amount of residual

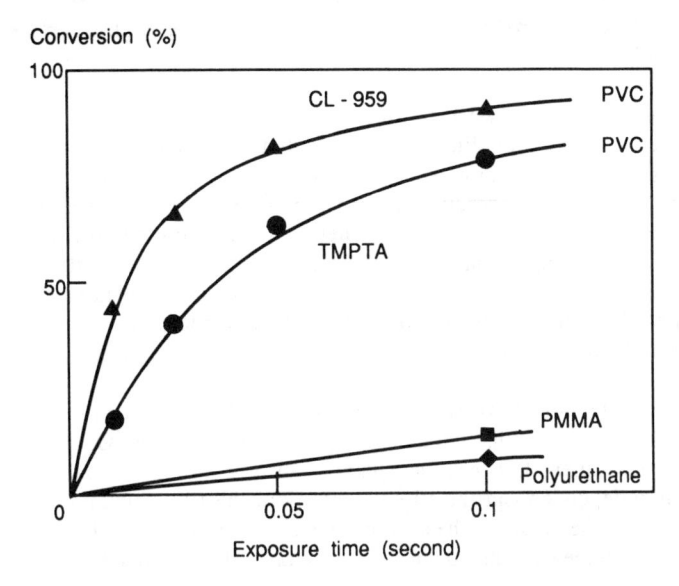

Fig. 7 : Photopolymerization profile of dry acrylic films irradiated in the presence of air (I = 500 mW cm^{-2})

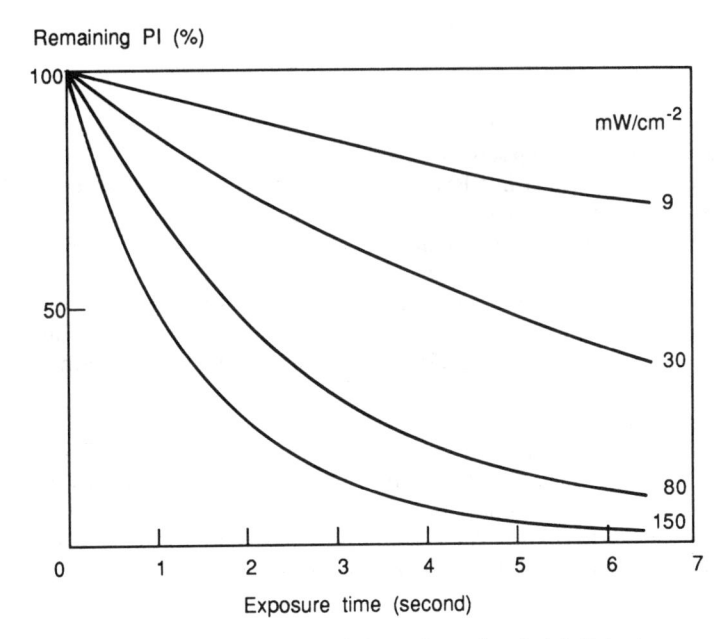

Fig. 8 : Influence of the light intensity on the photoinitiator loss profile in an acrylate monomer

photoinitiator in the final product, which is known to affect the longterm properties of UV-cured polymers.

Real time ultraviolet (RTUV) spectroscopy was used to monitor continuously the disappearance of a morpholino-type photoinitiator (Irgacure 369 from Ciba-Geigy) during the polymerization of multiacrylate monomers exposed to intense UV radiation. The photolysis reaction was found to follow a neat exponential law, $[PI]_t = [PI]_0 \exp(-kt)$, as shown by the kinetic profiles reported on Fig.8 for a sample exposed to UV radiation of various intensities. Such a first-order kinetic law was expected, since the rate of a photochemical reaction depends first of all on the probability of an incident photon to be absorbed by a photoinitiator molecule, and thus on the PI concentration.

The rate constant of PI loss was shown to be directly proportional to the light intensity[14], which indicates that the quantum yield of the reaction, $\emptyset_{-PI}$, i.e., the number of PI molecules destroyed per photon absorbed, remains constant. The $\emptyset_{-PI}$ value is independent on the PI concentration, the film thickness and O_2 concentration, but it decreases as the acrylate concentration is increased. This effect can be explained on the basis of a quenching reaction of the PI excited states by the acrylate monomer.

$$PI \xrightarrow{h\nu} PI^* \begin{array}{c} \searrow \\ M \nearrow \end{array} \begin{array}{c} \rightarrow PI + M^* \\ \hookrightarrow M + \text{heat} \end{array}$$

Such a quenching process occurs both in the liquid and in the solid state, and makes the initiation rate drop substantially. For example, the PI loss rate (R_{-PI}) was found to be 3 times greater in a PMMA film than in a triacrylate/PMMA (50/50) blend containing 6 mol L^{-1} of acrylate double bonds, and 7 times greater than in the neat triacrylate. Increasing the monomer concentration is thus causing both a decrease of the initiation rate and an increase in the polymerization rate, finally resulting in longer kinetic chains. This effect can be quantified by simply taking the ratio R_p/R_{-PI}, which is directly related to the kinetic chain length. Its value was found to increase from 150 for a 25 % diacrylate solution in chloroform ($[A]_0 = 1.7$ mol L^{-1}) to as much as 8000 for bulk PETIA ($[A]_0 = 12$ mol L^{-1})[14].

In radical-type UV-curable resins, the polymerization process develops usually much faster than the PI photolysis, so that there remain relatively large amounts of residual photoinitiator in the final product (up to 80 % of the original content). A quite different behavior was observed in cationic systems where PI was consumed much faster, thus leading to a cured polymer that contained essentially no residual photoinitiator[14].

Real-time ultraviolet spectroscopy has proved to be a most valuable technique to monitor the photolysis of initiators used in UV-curable resins. This very simple and reliable analytical method applies to any type of photoinitiators, radical or cationic, and can be carried out in solution, in bulk or in a polymer matrix. When employed in conjunction with RTIR spectroscopy, it should be most useful for assessing the performance of some new photoinitiators, and selecting the best candidate for the application considered.

4. CONCLUSION.

Radiation curing has now become a well accepted technology which has found its major openings in the coating industry and in photolithography, due to the rapidity and spatial selectivity of the process. The kinetics of such ultrafast polymerizations can be studied quantitatively by real time UV and IR spectroscopy. From the conversion vs. time profiles which are directly recorded in the millisecond timescale, the important kinetic parameters have been determined, and the effect of chemical and physical factors on the polymerization efficiency has been quantified for multiacrylate resins exposed to UV radiation or laser beams.

The reactivity of the acrylic monomers used as diluents in most UV-curable formulations can be greatly enhanced by introducing into the structural unit some functional groups which exhibit a strong hydrogen donor character, and are thus likely to favor chain transfer reactions. An additional interest of some of these new monomers is that they impart both hardness and flexibility to the UV-cured coating, thus providing scratch and impact resistance, two properties which are usually difficult to achieve at once. The main applications of these mono-acrylates are expected to be found in areas where fast and extensive polymerization is required in order to produce high-quality organic materials.

REFERENCES

1. C.G. Roffey, "Photopolymerization of Surface Coatings" Wiley & Sons, New York, 1982.
2. S.P. Pappas (ed.) "UV-Curing Science and Technology" Vol. 1 and 2, Techn. Market. Corp., Norwalk, 1978 and 1985.
3. L.F. Thompson, C.G. Wilson and J.M. Frechet (eds) "Material for Microlithography - Radiation Sensitive Polymers", Am. Chem. Soc. Washington DC, 1984.
4. K.K. Dietlieker and J.V. Crivello in "Chemistry and Technology UV & EB Formulation for Coatings Inks and Paints", P.K.I. Oldring (ed.), SITA Techn. London, 1991.
5. C.E. Hoyle and J.F. Kinstle (eds) "Radiation Curing of Polymeric Materials", Am. Chem.Soc. Washington DC, 1991.
6. C. Decker, Polym Photochem., 1983, 3, 131.
7. C. Decker, J.Polym.Sci. Polym. Chem. Ed., 1983, 21, 2451.
8. G.Y. Chen, Printed Circuits Fabrication, 1986, 1, 41.
9. G.S. Kumar and D.C. Neckers, Macromolecules, 1991, 24, 4322.
10. B.L. Booth, J. Appl. Photo. Eng., 1977, 3, 24.
11. C. Decker and K. Moussa, Makromol. Chem., 1988, 189, 2381.
12. C. Decker and K. Moussa, J. Coat. Techn., 1990, 62(786), 55.

13. C. Decker and K. Moussa, J. Polym. Sci., Polym. Chem. Ed., 1990, 28, 3429.
14. C. Decker, J.Polym. Sci. Polym. Chem. Ed., 1992, 30, 913.
15. C. Decker and K. Moussa, Macromolecules, 1989, 22, 4455.
16. C. Decker and K. Moussa, Makromol. Chem., 1990, 191, 963.
17. C. Decker and K. Moussa, Europ. Polym. J., 1991, 27, 403 and 881.
18. C. Decker and K. Moussa, Makromol. Chem Rapid Commun, 1990, 11, 159.
19. C. Decker and K. Moussa, Makromol. Chem., 1991, 192, 507.
20. R. Zertani and D. Mohr, Germ. Pat. DE 3.832.032 (1988).
21. C. Decker and K. Moussa, Proceed RadTech Conf. Boston, 1992, p.260.
22. C. Decker, Macromolecules, 1990, 23, 5217.
23. C. Decker in "Laser-Assisted Processing" Vol. 2, L. Laude (ed), SPIE, Washington DC, 1990, p. 50.
24. C. Decker and K. Moussa, Europ. Polym. J., 1990, 26, 393.
25. K. Moussa and C. Decker, Proceed. RadTech Conf. Boston 1992, p. 291.

Interactions of Polymer Surfaces with Ultraviolet Laser Pulses

R. Srinivasan

UV TECH ASSOCIATES, 2508 DUNNING DRIVE, YORKTOWN HEIGHTS, NY 10598, USA

1 INTRODUCTION

Over the past decade, the interaction of pulsed, ultraviolet laser radiation from an excimer laser (193nm, 248nm, 308nm or 351nm) with organic polymer surfaces has been the subject of intense research activity.[1,2] At power densities greater than 1 MW/cm^2 and using laser pulses of <1µs pulse width (FWHM), these interactions lead to photoablation or ablative photodecomposition (APD)[3,4] which results in the etching of the polymer surface and the explosive ejection of the decomposition products at supersonic velocities. The result is an etch pattern in the solid with a geometry that is defined by the light beam. The principal advantages in using ultraviolet laser radiation rather than visible or infrared laser radiation for this purpose lie in the precision (±2000 Å) with which the depth of the cut can be controlled and the lack of thermal damage to the substrate to a microscopic level. Within a period of a year after the first report, other groups confirmed these observations in several other polymers and at different wavelengths.[5-8] The possibility of extending the process to biological tissue was reported in 1983.[9]

Research devoted to understanding the science and technology behind UV laser ablation of polymers has grown at a surprising rate over the past decade.[1,2] By the use of excimer lasers of special design[10] as well as by elaborate arrangements for frequency-doubling and amplification, the widths of the pulses of use in the study of UV laser: polymer interactions have been extended from 0.3µs[11] to 160fs.[12,13] Pulses longer than 1µs have been generated by chopping the cw beam from an argon-ion laser which emitted UV laser radiation. The interaction of these pulses which were up to 1ms duration with organic polymers has recently been reported.[14] In discussing these interactions, it is relevant to consider the power density of the radiation at the polymer surface instead of the fluence as is commonly done. In Fig.1, a matrix of pulse widths and power

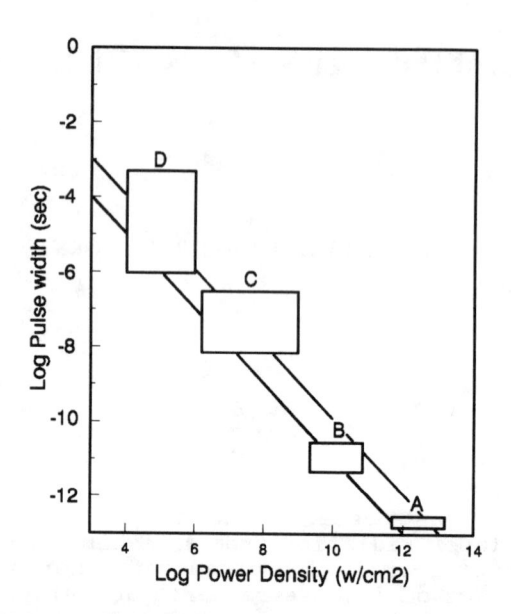

Figure 1 Matrix of power densities and pulse widths at which laser: polymer interactions have been studied. A. femtosecond pulses, B. picosecond pulses, C. nanosecond pulses, D. microsecond pulses. The diagonal lines correspond to constant fluences of 1J/cm^2 (upper) and 0.1J/cm^2 (lower) resp.

densities is displayed and the limits over which the investigations of UV laser polymer interactions have been carried out are marked.

2 Characteristics of UV Laser Ablation

Nearly all organic polymers show moderate to intense absorption in the ultraviolet region. These absorptions are usually ascribed to electronic transitions from a ground singlet to the first excited singlet states. The unique features in the UV laser ablation of polymers are encountered only in those wavelength regions in which such electronic absorptions exist.

The ablation of the surface of a polymer by a UV laser pulse is a function of the energy deposited in the solid in unit time. If a typical UV pulse has a full width at half-maximum (FWHM) of 20ns, an energy of 450mJ and the size of the beam at the polymer surface is 1.5 cm^2, the power density at the surface will be 1.5×10^7 W/cm^2. When this pulse strikes the surface a loud audible report will be heard and, depending upon the wavelength, 0.01-0.1μm of the material would have been etched away with a geometry that is defined by the light beam. If this experiment is performed in air, a bright plume will be ejected from the surface and will extend to a few millimeters. A more detailed description of this phenomenon can be found in the Section 4.

Typically, UV laser ablation is carried out with a succession of pulses. The etching of the surface is a linear function of the number of pulses when the polymer is a strong absorber at the laser wavelength. In the case of a weak absorber such as polymethyl methacrylate (PMMA) at 248nm, a phenomenon which has been termed "incubation"[15] is observed. It results in the first few pulses not causing any etching at all and even giving rise to a raised surface. But after a few pulses - the exact number depending upon the fluence, the wavelength and the absorption of the polymer - the system settles down to a constant etch depth per pulse. Research devoted to the establishment of the cause of this "incubation" effect [16-18] suggests that more than one reason may be important. The value of the etch depth per pulse is usually averaged over hundreds of pulses in order to minimize the uncertainties that are introduced in the measurement of the etch depth as well as the incubation effects where they exist. Average values for the etch depth/pulse are reproducible to within the uncertainties in the measurement of the fluence provided the absorption characteristics of the polymer are well-controlled.

The use of UV laser ablation to etch polymer surfaces has become an established technology.[19] The potential for such use was an important factor in the progress that has been made in the investigation of this phenomenon.[2] With this perspective in mind, nearly all of the etch data have been presented in the literature as a plot of etch depth/pulse vs. log fluence. It is intuitive that the interaction of photons with the polymer should involve the power density (power/ unit area) rather than the fluence. Discussions based on the fluence are acceptable only so long as the pulse width is constant. The major portion of the data that were published until the last few years involved the use of commercial excimer lasers whose pulse widths fall within a narrow range of 15-35ns (FWHM). Therefore, the use of fluence rather than the power density as the laser parameter that controlled the etch depth (at a given wavelength and for a particular polymer) was acceptable.

3 Analysis of the Published Data

Published data on UV laser ablation can be sorted out in terms of a simple representation of the phenomenon as in Fig. 2. The three cartoons in this figure show the three successive steps of absorption of the photon, break-up of the structure of the solid and the ablation of the

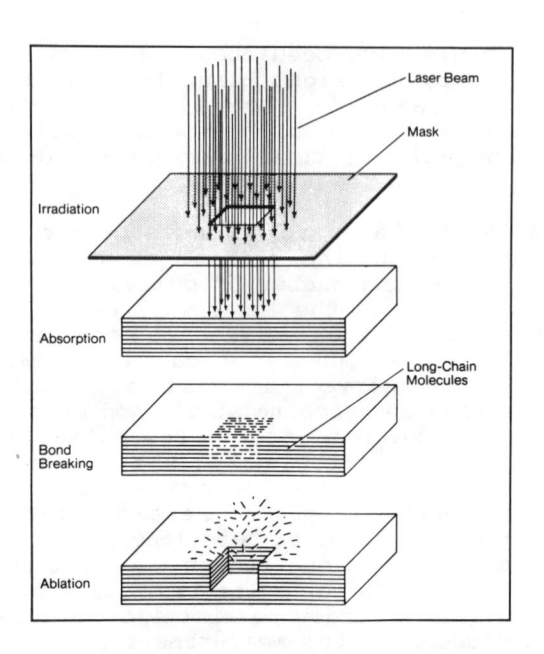

Figure 2 Hypothetical steps in the interaction of a laser beam with a polymer surface. Top. The laser radiation which is defined by a mask is absorbed. Middle. Chemical bonds in the polymer are broken by the photon energy. Bottom. The products ablate at supersonic velocities to leave an etched sample.

products. These steps are presumed to be the result of a single laser pulse interacting with the surface. The published data fall into three categories which correspond to averaging the results of many pulses, or studying the overall result of a single pulse, or probing the events during a single pulse. Data on the etch depth/ pulse which were discussed in Section 2 represent averaging the results of many hundreds of pulses as pointed out already. Product analyses which are carried out by chemical methods also fall in this category. Data that are obtained by analyzing the results from a single pulse fall into the second category. These data can be looked upon as a time-average since the system is usually sampled at a time interval that is long (microseconds) when compared to the duration of a single laser pulse. Such measurements are made by physical methods including mass spectometry and optical spectroscopy. An extensive review of the research along these lines has been published.[1] The third category consists of results that are obtained by probing the system at intervals that are small or are of the same order as the pulse width. A significant amount of research work in recent years has been devoted to this approach.

Interactions of Polymer Surfaces with Ultraviolet Laser Pulses 51

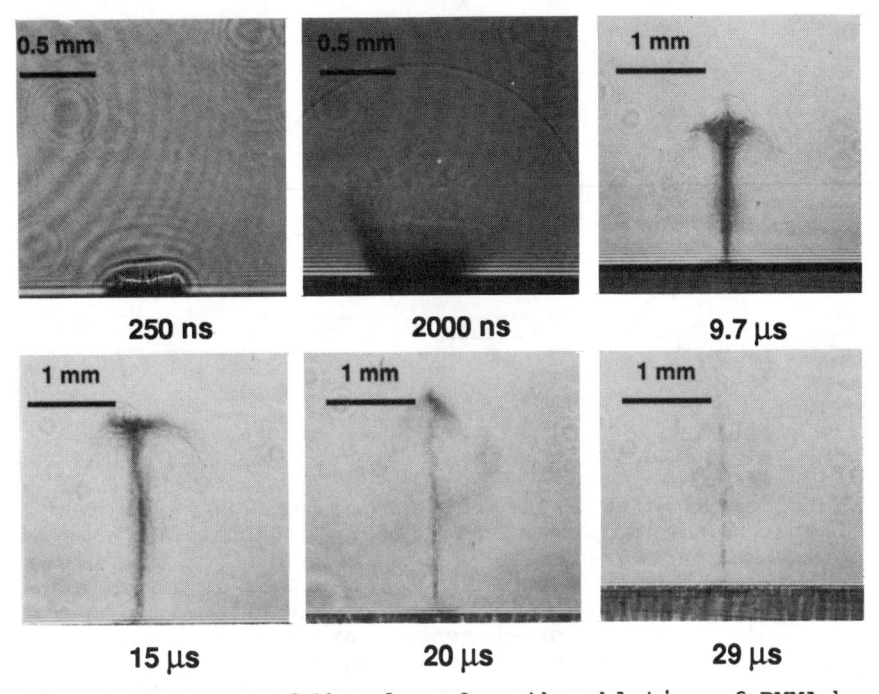

<u>Figure 3</u> Course of the plume from the ablation of PMMA by 248nm laser pulses shown as a function of time. Fluence: 1.34 J/cm^2. (Ref. 16)

4 Rapid Probing of Ablation

When a polymer film that is a few microns thick is glued to a piezo-electric sensor which consists of a polyvinylidene fluoride film with evaporated metal electrodes on its surfaces, the ablation of the polymer surface by a UV laser pulse registers an electrical signal between the metal film electrodes. It was first shown in 1986[20] that this signal was due to the stress wave that travels through the thickness of the polymer and is picked up by the piezo-electric sensor. It is the reaction to the ejection of the ablation products from the surface of the polymer. A typical set of results from these experiments shows that the stress wave begins to grow within a few nanoseconds after the start of the laser pulse and has about the same duration as the laser pulse itself. This suggested that ablation was a fast process, but the speed at which the products of the ablation were transported away from the polymer surface was not established until several years later when many groups, nearly simultaneously, showed by fast photographic methods that the stream of products from the ablated surface lasted over many tens of microseconds. In one set of experiments the emissive species were photographed by fast framing cameras.[21] In several other investigations, the blast front and the solid products that were ablated were

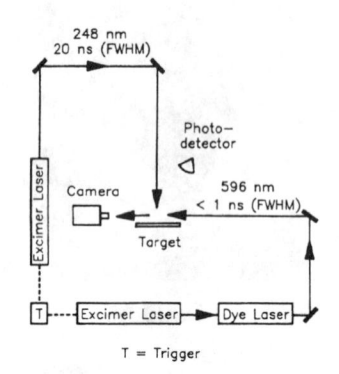

Figure 4 Schematic diagram of experiment as set up for photographs in Fig. 3 (Ref. 16)

photographed with a conventional camera using a fast, visible laser pulse as the probe.[22-24] Figure 3 shows a series of photographs in which the course of the plume from the ablation of PMMA by 248nm pulses is shown as a function time. These experiments were conducted in an air atmosphere in an experimental arrangement which is illustrated in Figure 4. The gas stream of products which issues at supersonic velocities has assumed a hemispherical envelope as it encounters the air atmosphere but the solid particulate matter which travels at subsonic velocity maintains its normal direction to the polymer surface. The fate of the plume at relatively long time intervals and at several millimeters distance from the polymer surface has recently been made clear[25] by the use of pulsed schlieren photography. The plume detaches itself from the surface at times of the order of 1ms. Analysis of the velocity measurement in the light of point explosion theory suggests that to drive the plume chemical reaction between the products of ablation and air may provide additional energy over that provided by the photons in the case of some polymers. This driving force is also dependent on the fluence and the atmosphere above the polymer surface.

5 Future Directions

The general features of the UV laser ablation of polymers seem clear now but a detailed understanding of the phenomenon has not been reached. It is necessary to probe the system at time intervals of 1 - 1000ns and at distances of less than a few millimeters from the surface. If optical spectroscopy, laser-induced-fluorescence and mass spectrometry can be carried out in this early region in time and space, it is very likely that useful insights into the mechanism of the ablation process can be gained. In parallel with these investigations, it is also necessary to record the changes in the polymer in the thin layer of material that lies just below the new surface that is created by the ablation pulse. If these data can be obtained for several different polymers at the three principal excimer wavelengths, our understanding of UV laser ablation will be certain to be broadly expanded.

6 References

1. R. Srinivasan and B. Braren, Chem. Rev., 1989, 89, 1303.
2. P.E. Dyer, "Laser Ablation of Polymers" in 'Photochemical Processing of Electronic Materials', Ed. I.W. Boyd, Academic Press, London, 1992, p.359.
3. R. Srinivasan and V. Mayne - Banton, Appl. Phys. Lett., 1982, 41, 576.
4. R. Srinivasan and W.J. Leigh, J. Am. Chem. Soc., 1982, 104, 6784.
5. T.F. Deutsch and M.W. Geis, J. Appl. Phys., 1983, 54, 7201.
6. M.W. Geis, J.N. Randall, T.F. Deutsch, N.N. Efremow, J.P. Donnelly and J.D. Woodhouse, J. Vac. Sci. Tech., 1983, B1, 1178.
7. Y. Kawamura, K. Toyoda, and S. Namba, Appl. Phys. Lett., 1982, 40, 374.
8. J.E. Andrew, P.E. Dyer, D. Forster and P.H. Dey, Appl. Phys. Lett., 1983, 43, 717.
9. S.T. Trokel, R. Srinivasan and B. Braren, Am. J. Ophthalmol., 1983, 96, 710.
10. R.S. Taylor and K.E. Leopold, J. Appl. Phys., 1989, 65, 22.
11. R.S. Taylor, D.L. Singleton and G. Paraskevopoulos, Appl. Phys. Lett., 1987, 50, 1779.
12. S. Kuper and M. Stuke, Appl. Phys. Lett., 1989, 54, 4.
13. M.C. Chuang and A.C. Tam, J. Appl. Phys., 1989, 65, 2591.
14. R. Srinivasan, J. Appl. Phys., 1992, 70, August.
15. E. Sutcliffe and R. Srinivasan, J. Appl. Phys., 1986, 68, 3315.
16. R. Srinivasan, B. Braren and K.G. Casey, J. Appl. Phys., 1990, 68, 1842.
17. S. Kuper and M. Stuke, Appl. Phys., 1988, B44, 199.
18. S. Kuper and M. Stuke, Appl. Phys., 1989, A49, 211.
19. F. Bachmann, Chemtronics, 1989, 4, 149.
20. P.E. Dyer and R. Srinivasan, Appl. Phys. Lett., 1986, 48, 445.
21. P.E. Dyer and J. Sidhu, J. Appl. Phys., 1988, 64, 4657.
22. C.A. Puliafito, D. Stern, R.R. Krueger and E.R. Mandel, Arch. Ophthalmol., 1987, 105, 1255.
23. R. Srinivasan, B. Braren, K.G. Casey and M. Yeh, Appl. Phys. Lett., 1989, 55, 2790.
24. R. Srinivasan, K.G. Casey, B. Braren and M. Yeh, J. Appl. Phys., 1990, 67, 1604.
25. P.L.G. Ventzek, R.M. Gilgenbach, J.A. Sell and D.M. Heffelfinger, J. Appl. Phys., 1990, 68, 965.

Excimer and CO_2 Laser Ablation of Organic Polymers

G. A. Oldershaw

SCHOOL OF CHEMISTRY, UNIVERSITY OF HULL, HULL HU6 7RX, UK

1 INTRODUCTION

The ablation of an organic polymer by pulsed ultraviolet radiation from an ArF laser was first reported in 1982 by Srinivasan and co-workers[1,2]. In this process etching of the polymer occurs at the site of irradiation and material is forcibly ejected from the surface. Subsequently there has been increasing interest in, and work on, excimer laser ablation, sustained by actual and potential applications in micromachining and lithography. Similar work on biological materials has applications in laser surgery. Several reviews, covering the ablation of a range of materials with excimer lasers of various output wavelengths, have been published[3-6].

Two features of excimer laser ablation have been observed generally. Firstly, the interaction between the ultraviolet radiation and the polymer is very localised, so that patterns of high definition and controlled depth can be etched in the surface. Because of the strong absorption of many polymers at 193 nm, the output wavelength of the ArF laser, the depth of the etch for a single laser pulse is very small, usually less than 1 μm. Secondly there is a threshold laser fluence below which etching, while detectable in some cases, is insignificant. For a limited range of fluence above threshold the depth of the etched hole l is given approximately by[7]

$$l = k_e^{-1} \ln(F/F_T) \qquad (1)$$

where F is the fluence, F_T the threshold fluence and k_e is the effective absorption coefficient for the laser radiation. The parameters F_T and k_e depend on the polymer and on the output wavelength of the laser. An example of the dependence of l on F is shown in Figure 1.

Measurements of the thermal loading of thin polymer films irradiated with excimer lasers have shown that below threshold essentially all the absorbed energy

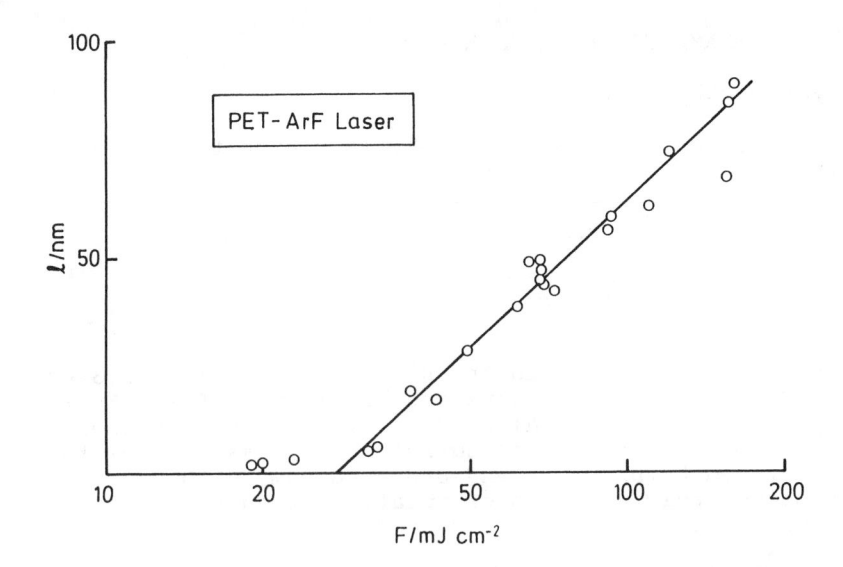

Figure 1 Etch depth per pulse as a function of fluence for ArF laser ablation of PET. Reprinted, with permission, from ref. 17. Copyright (1991) American Chemical Society.

appears as heat[8]. This implies that a substantial transient temperature rise occurs in the surface layer of the excimer-irradiated polymer, owing to the small absorption depth. Above threshold the thermal loading of the film is independent of the fluence as excess absorbed energy is carried off by the ablated material. Discussion about the mechanism of excimer laser ablation centres on the question of whether decomposition occurs by direct photolysis or by a photothermal mechanism, in which the initial electronic excitation is rapidly degraded to heat with subsequent thermolysis of the polymer. The timescale of the ablation process has been established by photoacoustic measurements of stress waves in excimer laser-irradiated films[9,10]. These show that for fluences above threshold ablation occurs rapidly, starting within the lifetime (~ 20 ns) of a typical laser pulse.

While there is a considerable literature on excimer laser ablation, only a limited amount of work on ablation of polymers by pulsed infrared radiation from TEA CO_2 lasers has been reported[11,12].

This article is concerned with studies of ablation of three particular materials, polyetheretherketone (PEEK), a PEEK-carbon fibre composite, and polyethylene terephthalate, using both excimer and CO_2 lasers.

2 EXAMPLES OF LASER ABLATION

Polyetherether ketone

Polyetheretherketone (PEEK) is a high temperature thermoplastic with the structure

PEEK absorbs strongly in the ultraviolet at both 193 nm and 308 nm, the output wavelengths of the ArF and XeCl lasers, and experimental studies of etching have been carried out at both wavelengths[13-15]. Films of PEEK have been etched both in air and under vacuum, with gas chromatographic analysis of volatile products in the latter case.

PEEK is readily etched by the ArF laser (193 nm) with a threshold fluence for ablation of about 28 mJ cm^{-2}. The major gaseous ablation products are carbon monoxide and ethyne, with smaller quantities of butadiyne (C_4H_2) and traces of other products.

Ablation of PEEK in air with the 308 nm XeCl laser also produces high quality etching with well-defined edges. An example of an etch crater produced by multiple-pulse exposure is shown in figure 2a. Particulate debris can be observed outside the rim of the crater (fig 2b), which gave the visual appearance of a sooty ring around the etch site, and there is some debris on the floor of the crater. Etching under vacuum produces no observable debris and no sooty ring. This may indicate that debris is produced by combustion in air of ablated material, or that, under vacuum, debris can escape from the vicinity of the etch site.

As in the case of ArF laser ablation, the main gaseous products of the XeCl laser ablation are carbon monoxide and ethyne, and methane, butadiyne and small quantities of benzene and other C_3 and C_4 hydrocarbons are also produced. Both the total quantity and composition of the gaseous products are strongly dependent on fluence. Figure 3 shows the yields of the two major products CO and C_2H_2 relative to the number of repeat units $C_{19}H_{12}O_3$ removed from the polymer by ablation. Although the threshold fluence for removal of polymer is 70 mJ cm^{-2}, no significant production of gaseous species occurs below 100 mJ cm^{-2}. Thus at fluences just above the threshold, the ablated material consists of involatile species of relatively high molecular weight, whereas at higher fluences the extent of fragmentation of the ablation products increases. At sufficiently high fluence 3 moles of CO per PEEK repeat

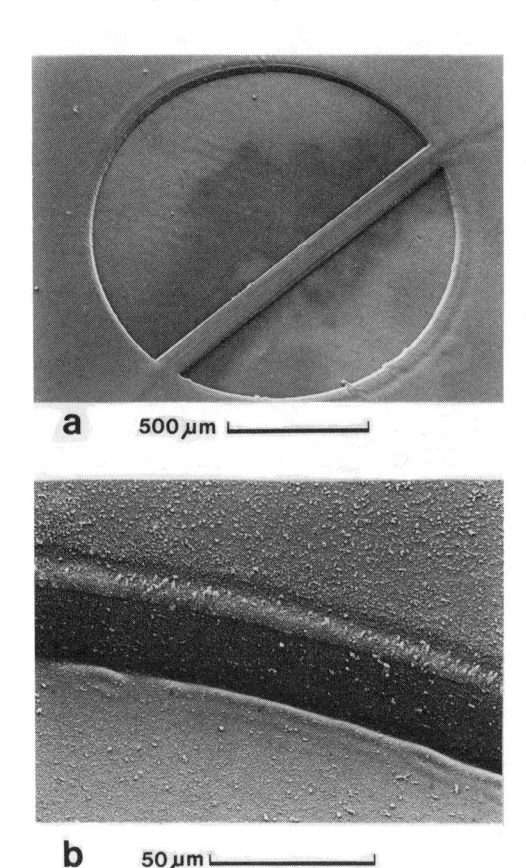

Figure 2 Scanning electron micrographs of XeCl laser ablated PEEK film. (a) Complete etch site; the line corresponds to a wire placed on the aperture to assist focusing. (b) Detail of crater wall showing particulate debris. Reproduced, with permission, from ref. 14.

unit are produced, corresponding to conversion of all the oxygen in the ablated polymer to CO. The CO therefore originates both from the carbonyl group and the ether linkages in the polymer, illustrating the extensive disruption of the structure that occurs at high fluences. Fragmentation of the aromatic rings gives C_2H_2, C_4H_2 and a substantial quantity of carbon-rich involatile material undetected by the analytical method employed.

Etching of PEEK with pulsed infrared radiation from the CO_2 laser is difficult because of the relatively weak absorption in the region of the output wavelengths of the CO_2 laser. Very high fluences are required and the etching is of poor quality. When the laser radiation is only weakly absorbed, there is a large penetration depth, cooling by conduction is slow, and melting and poor etching results. By contrast, the strong absorption of

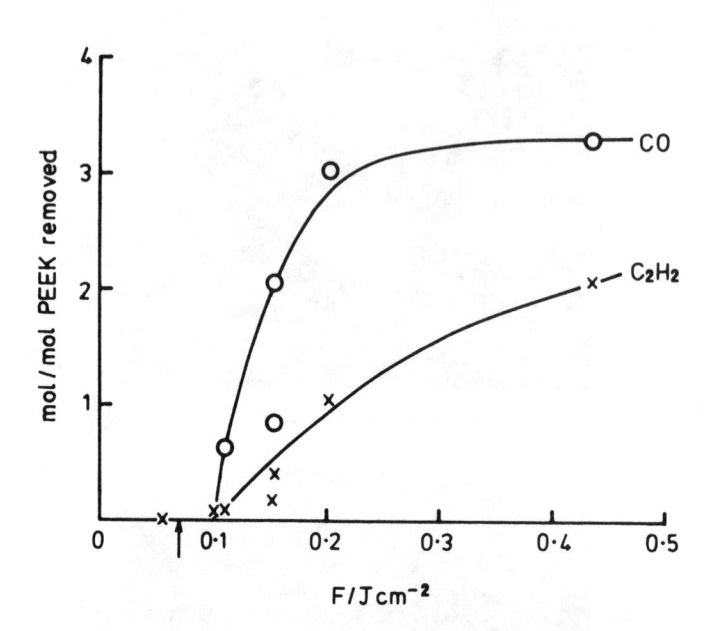

Figure 3 Major products of the ablation of PEEK as a function of fluence. The arrow shows the ablation threshold. Reproduced, with permission, from ref. 14.

PEEK at 193 nm and at 308 nm gives rise to very small absorption depths and high quality etching by the excimer lasers.

PEEK/carbon fibre composite

The extension of work on laser ablation to synthetic composite materials is clearly of interest and an example is provided by the XeCl laser ablation of a PEEK-carbon fibre composite (APC-2, ICI plc)[16]. This material consists of carbon fibres of diameter 6-9 μm embedded in a PEEK matrix. Measurements of the etch rate of the composite at different fluences are shown in Figure 4 where they are compared with data for PEEK. As would be anticipated, the presence of the graphitic fibres in the composite raises the threshold for etching, to about 420 mJ cm^{-2}. For fluences between 70 and 400 mJ cm^{-2} selective removal of the PEEK matrix in the surface layer of the composite occurs, without significant etching of the carbon fibres, revealing the underlying fibre layout.

Above 420 mJ cm^{-2} the PEEK matrix ablates readily and the composite etch rate is governed by that of the carbon fibres. The etch depth can be successfully modelled on the assumption that the UV laser radiation heats the graphite fibres, which are than vaporised. Two limiting cases can be identified. At low irradiance levels the surface temperature T_s is determined by

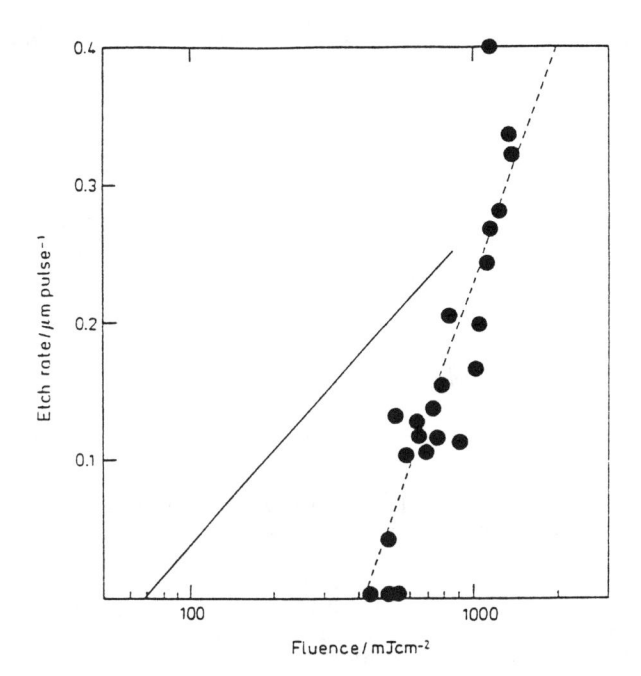

Figure 4 Etch rate for PEEK-carbon fibre composite as a function of XeCl laser fluence. The solid line summarises the etch rate for PEEK alone. Reproduced, with permission, from ref. 16.

conduction of heat from the surface, and the rate of surface recession due to the evaporation of any particular gaseous carbon species can be calculated from its pressure in equilibrium with graphite at T_s. In this limit cooling due to the vaporisation is neglected. At the relevant temperatures the main gaseous species involved is C_3 with contributions from C_1, C_2 and C_4. Etch depths calculated in this way are shown as curve A in Figure 5 and predict a threshold close to the observed value. At sufficiently high irradiance the surface temperature is limited by cooling due to vaporisation, and the etch rate at any particular fluence can be calculated from the enthalpy of vaporisation. This limiting case is shown by line B, and the experimental etch rate adjusted for the proportion of carbon fibre in the composite is line C. The model based on thermal evaporation of graphite at temperatures in the range 4000 to 6000 K thus provides an adequate representation of the composite etch rates.

Polyethylene terephthalate

CO_2 laser ablation. Polyethylene terephthalate (PET), $(-OCOC_6H_4COOC_2H_4-)_n$, absorbs strongly at 193 nm and adequately at 308 nm and is easily ablated by ArF or XeCl laser radiation. Etching of good quality can also

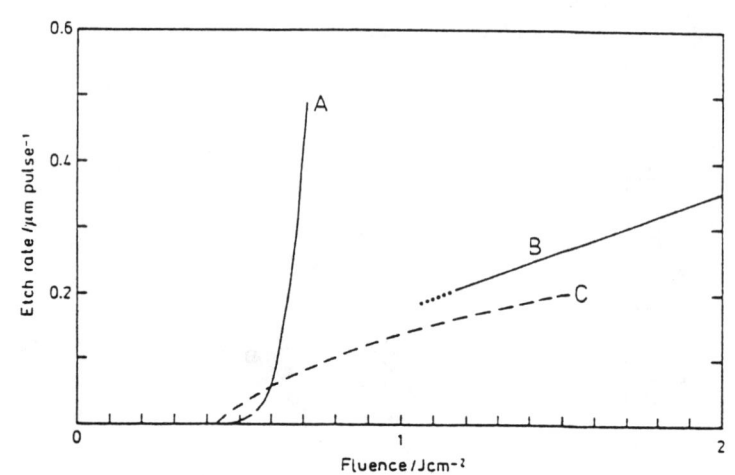

Figure 5 Calculated and measured etch rates. Reproduced, with permission, from ref. 16.

be achieved with pulsed infrared radiation from a CO_2 laser provided that one of the more strongly absorbed CO_2 lines is used[12]. This is of interest since it shows that thermal decomposition induced by infrared absorption can lead to rapid ablation.

The timing of the ablation process has been examined by recording photoacoustic transients during laser irradiation. Stress waves generated in the polymer film are of two main types[10]. Heating of the film by the laser pulse gives rise to a bipolar thermoelastic stress wave, and, in the case of irradiation above the threshold fluence, the ablation process leads to a compressive stress as material leaves the surface. Figure 6 shows

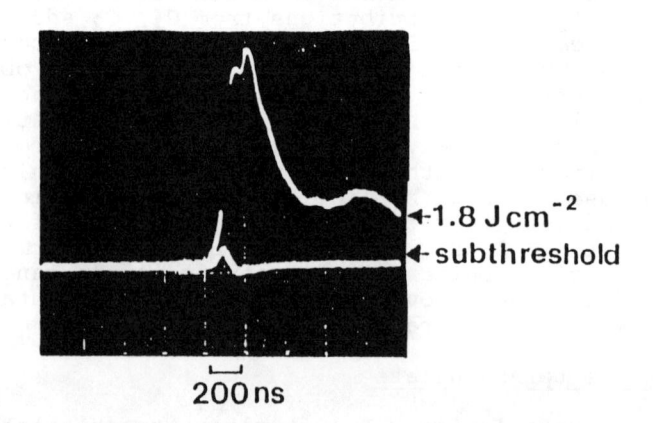

Figure 6 Acoustic signals for pulsed CO_2 laser irradiation of PET. Reproduced, with permission, from ref. 12.

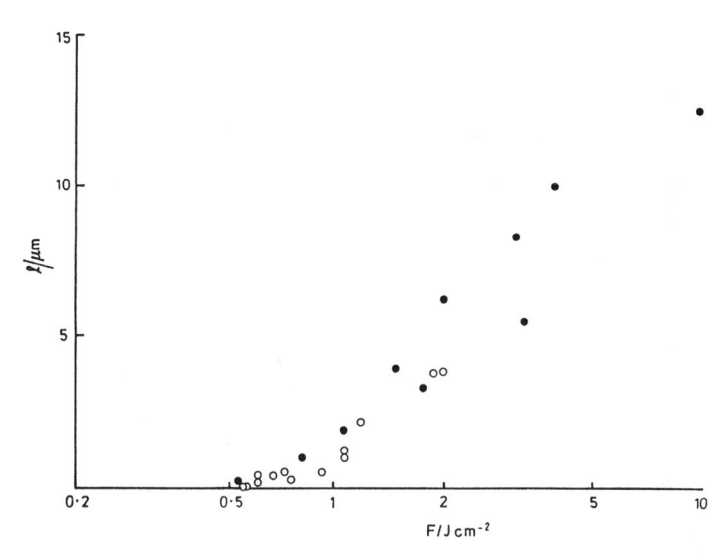

[Graph: Etch depth ℓ/μm vs F/J cm⁻²]

Figure 7 Etch depth per pulse as a function of fluence for CO₂ laser ablation of PET. 9R42 line. Reproduced, with permission, from ref. 12.

signals recorded for a PET film irradiated with the 9R40 CO_2 line at 1090 cm^{-1}. The smaller signal records the bipolar thermoelastic stress wave produced at a fluence below threshold, and the larger signal is the result of irradiation at 1.8 J cm^{-2}, which causes ablation. In this case the initial positive part of a thermoelastic stress is followed by another positive signal due to ablation. Ablation starts within about 100 ns of the beginning of the ~200 ns laser pulse.

Measurements of etch depths show that the dependence on fluence is similar in form to that observed in excimer laser ablation, with some 'tailing' in the threshold region (Figure 7).

The major gaseous products of CO_2 laser ablation of PET are CO and CO_2, and CH_4, C_2H_2, C_2H_4, C_6H_6, CH_3CHO, C_4H_2 and C_4H_4 are also formed. As would be expected, the products are qualitatively similar to those found in the pyrolysis of PET. The production of CO and CO_2 as a function of fluence is compared with the removal of PET in Figure 8. There is a threshold for generation of the gaseous products of about 2 J cm^{-2}, more than double the threshold for removal of polymer. Thus at fluences in the threshold region relatively high molecular weight material is ejected by ablation, but at higher fluences more extensive decomposition to gaseous species occurs.

Excimer laser ablation. There has been speculation about the mechanism of excimer laser ablation of PET ever since the early experimental investigations involving

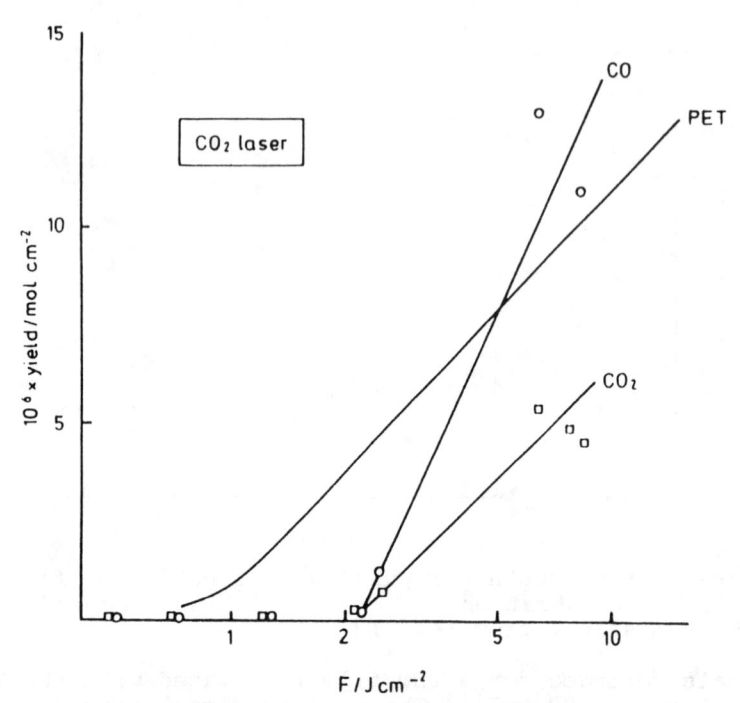

Figure 8 PET removed and CO and CO_2 production in CO_2 laser ablation. 9R42 line. Reproduced, with permission, from ref. 12.

this polymer[1,7]. The point at issue is the relative importance of direct photolysis and of thermolysis following the rapid degradation of the initial electronic excitation to heat, the photothermal mechanism. More recent work has a bearing on this question.

The gaseous products of ablation of PET by both the XeCl and ArF lasers are CO, CO_2, CH_4, C_2H_2, C_2H_4, C_4H_2, C_4H_4, C_6H_6 and CH_3CHO, and the composition of the products is broadly similar for CO_2, XeCl and ArF laser ablation[17]. This indicates that in excimer laser ablation at fluences well above threshold thermal decomposition of PET occurs, as is to be expected from the substantial surface temperature rises implied by thermal loading experiments[8]. It does not, however, answer the question of whether the initial decomposition which causes ablation at fluences close to threshold is due to direct photolysis or thermolysis.

Figure 9 shows the etch curve for XeCl laser ablation of PET which has been preheated by a pulse from a CO_2 laser, compared with that for ablation of PET initially at room temperature[18]. The preheated film has a significantly lower threshold fluence but its effective absorption coefficient is unchanged. The lowering of the

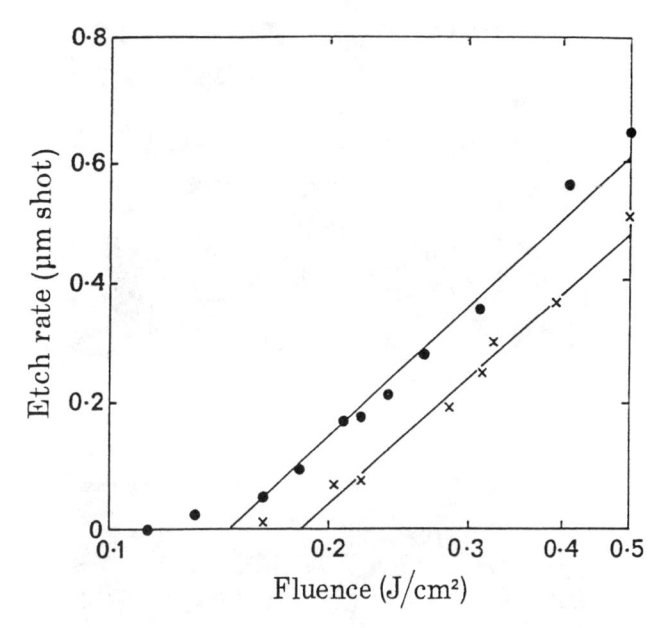

Figure 9 Etch rate of PET as a function of XeCl laser fluence; ●, film preheated by CO_2 laser pulse; ×, film at room temperature. Reproduced, with permission, from ref. 18.

threshold by heating supports a thermal mechanism for the ablation.

Further evidence of the importance of thermal factors in the XeCl laser ablation is provided by photoacoustic measurements on XeCl laser-irradiated PET. Figure 10A shows the transducer signal recorded for irradiation at a fluence well below the threshold. The bipolar thermoelastic stress wave agrees with that calculated on the assumption that the absorption of UV radiation produces electronic excitation which is instantaneously converted to heat. Relaxation of the absorbed energy to heat therefore occurs rapidly, in a time too short for observation with nanosecond resolution. The transducer response for irradiation at a fluence above threshold is shown in Figure 10B. The initial part of the thermoelastic stress wave is followed by a positive compressive signal due to ablation which outweighs the later part of the thermoelastic stress. Relaxation of the initial electronic excitation to heat therefore precedes ablation.

These experimental observations favouring a photothermal mechanism for the XeCl laser ablation are supported by a model which relates the threshold fluence to the rate of thermal decomposition of PET[19]. The initial step in the thermolysis is believed to be the

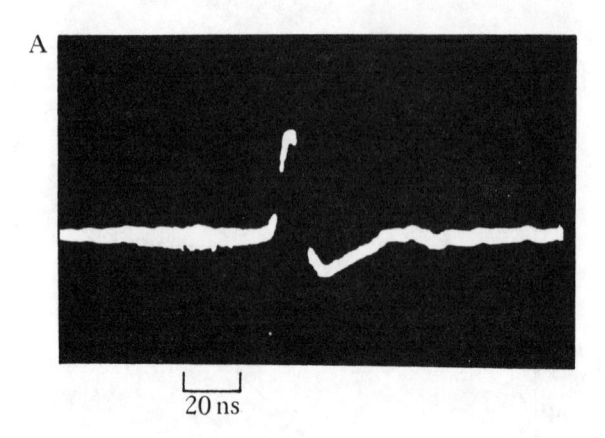

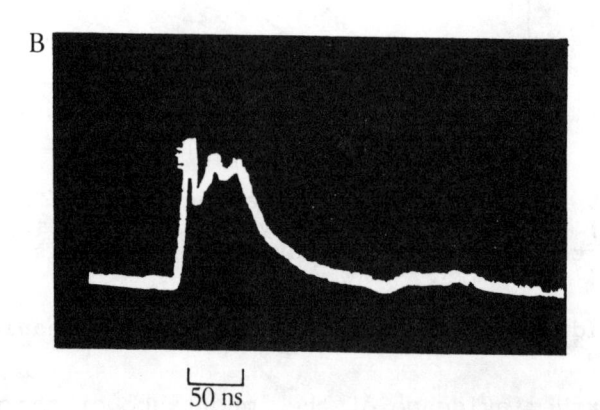

Figure 10 Transducer responses in the XeCl laser irradiation of PET. A, fluence = 0.06 J cm^{-2}; B, fluence = 0.24 J cm^{-2}. Reprinted, with permission, from ref. 17. Copyright (1991) American Chemical Society.

decomposition of the ester linkage to carboxylic acid and olefinic groups[20,21]:

$-C_6H_4COOCH_2CH_2OCOC_6H_4-$

$\longrightarrow$ $-C_6H_4COOH + CH_2=CHOCOC_6H_4-$ (2)

This process involves a significant increase in volume and the Arrhenius parameters are approximately[12,21] $A = 1.4 \times 10^{12}$ s^{-1} and $E_A = 196$ kJ mol^{-1}.

The model assumes that relaxation of the initial electronic excitation to heat is rapid compared both with conductive heat transfer and with the thermal decomposition of the polymer. The temperature of the surface layer of PET is calculated as a function of time, allowing for conduction of heat away from the surface,

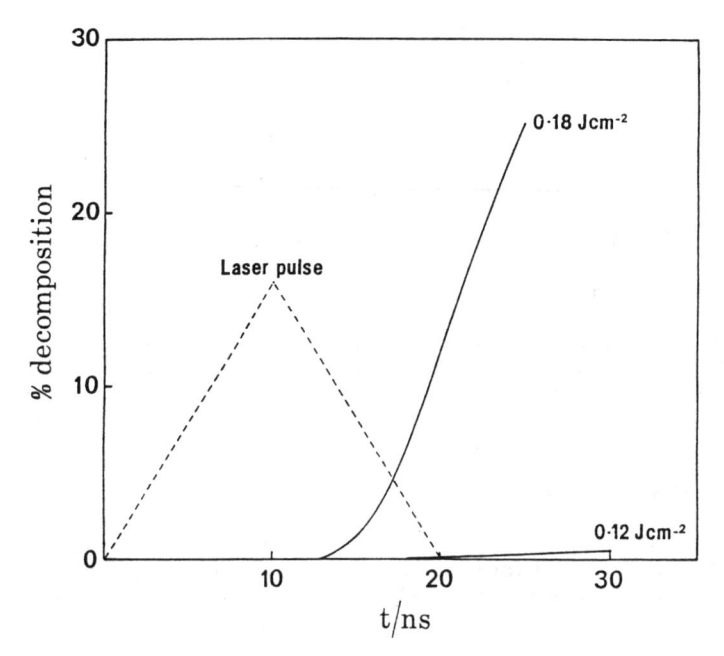

Excimer and CO₂ Laser Ablation of Organic Polymers

Figure 11 Effect of laser fluence on the calculated decomposition of PET during XeCl laser irradiation. Reproduced, with permission, from ref. 19.

and the extent of decomposition is computed from the temperature. Results for a typical 20 ns laser pulse at two different fluences are shown in Figure 11. The strong dependence of the extent of decomposition on laser fluence accounts for the existence of a threshold fluence for ablation. If the arbitrary but reasonable assumption is made that ablation occurs if there is at least 10% decomposition of PET by the end of the laser pulse, a threshold of 0.17 J cm^{-2} is predicted, as observed experimentally. Moreover, similar calculations for the CO₂ laser, where ablation is thermal in origin, using the same criterion for ablation, give a threshold which is also in agreement with experiment, as shown in Table 1.

Table 1 Calculated ablation thresholds

Laser	$\underline{k}e$/cm^{-1}	F_T/J cm^{-2} calc	exp
CO₂ (9.2 μm)	4×10^3	0.73	0.6
XeCl (308 nm)	2.2×10^4	0.17	0.17
KrF (248 nm)	1.2×10^5	0.037	0.030
ArF (193 nm)	2.0×10^5	0.026	0.028

Thus, while some direct photolysis of PET is expected to occur as a result of XeCl laser irradiation, ablation occurs mainly as a result of rapid relaxation of electronic energy to heat followed by thermal decomposition. The agreement of the calculated and observed thresholds for the other excimer lasers is also strong evidence for the importance of thermolysis in ablation in these cases, although the possibility of some contribution from photolysis cannot be excluded.

3 CONCLUSION

In the examples of laser ablation described in this article the mechanism of ablation is reasonably well-established. Much remains to be done in exploring the effects of both excimer and CO_2 laser irradiation on a wider range of polymeric materials.

ACKNOWLEDGEMENTS

Much of the work described in this article was carried out in collaboration with P.E. Dyer, D. Schudel and J. Sidhu. I am grateful to them for the provision of published and unpublished information.

REFERENCES

1. R. Srinivasan and V. Mayne-Banton, Appl. Phys. Letters, 1982, 41, 756.
2. R. Srinivasan and W.J. Leigh, J. Am. Chem. Soc., 1982, 104, 6784.
3. R. Srinivasan, Science, 1986, 234, 559.
4. R. Srinivasan and B. Braren, Chem. Rev., 1989, 89, 1303.
5. S. Lazare and V. Granier, Laser Chem., 1989, 10, 25.
6. P.E. Dyer, 'Photochemical Processing of Electronic Materials', ed. I.W. Boyd and R.B. Jackman, Academic Press, London, 1992, Ch. 14.
7. J.E. Andrew, P.E. Dyer, D. Forster and P.H. Key, Appl. Phys. Letters, 1983, 43, 717.
8. P.E. Dyer and J. Sidhu, J. Appl. Phys., 1985, 57, 1420.
9. P.E. Dyer and R. Srinivasan, Appl. Phys. Letters, 1986, 48, 445.
10. P.E. Dyer, 'Photoacoustic and Photothermal Phenomena', ed. P. Hess and J. Petzl, Springer, Berlin, 1988, p.164 .
11. J.H. Brannon and J.R. Lankard, Appl. Phys. Letters, 1986, 48, 1226.
12. P.E. Dyer, G.A. Oldershaw and J. Sidhu, Appl. Phys. B, 1989, 48, 489.
13. D. Schudel, Ph.D. Thesis, University of Hull, 1991.
14. P.E. Dyer, G.A. Oldershaw and D. Schudel, Appl. Phys. B, 1990, 51, 314.
15. E. Occiello, F. Garbassi and V. Malatesta, Angew. Makro. Chem., 1989, 169, 143.

16. P.E. Dyer, S.T. Lau, G.A. Oldershaw and D. Schudel, *J. Mater. Res.*, 1992, **7**, 1152.
17. P.E. Dyer, G.A. Oldershaw and J. Sidhu, *J. Phys. Chem.*, 1991, **95**, 10004.
18. R.K. Al-Dhahir, P.E. Dyer, J. Sidhu, C. Foulkes-Williams and G.A. Oldershaw, *Appl. Phys. B*, 1989, **49**, 435.
19. G.A. Oldershaw, *Chem. Phys. Letters*, 1991, **186**, 23.
20. M.E. Bednas, M. Day, K. Ho, R. Sander and D.M. Wiles, *J. Appl. Polymer Sci.*, 1981, **26**, 277.
21. L.H. Buxbaum, *Angew. Chemie, Intl. Ed.*, 1968, **7**, 182.

Photophysical Studies of Gelation and Cure in Polymeric Systems

R. A. Pethrick,[1*] B. Wandelt,[2] D. J. S. Birch,[3] R. E. Imhof,[3] and S. Radhakrishnan[4]

[1] DEPARTMENT OF PURE AND APPLIED CHEMISTRY, UNIVERSITY OF STRATHCLYDE, 295 CATHEDRAL STREET, GLASGOW GI IXL, UK
[2] INSTITUTE OF POLYMER SCIENCE, TECHNICAL UNIVERSITY, ZWIRKI 36, LODZ, POLAND
[3] DEPARTMENT OF PHYSICS AND APPLIED PHYSICS, UNIVERSITY OF STRATHCLYDE, GLASGOW GI IXL, UK
[4] POLYMER SCIENCE AND ENGINEERING, NATIONAL LABORATORY, PUNE 411 008, INDIA

1 ABSTRACT

Photophysical measurements of the gel state of isotactic polystyrene [iPs] illustrate the way in which measurements performed on a time scale of 10^{-9} sec. can be used to monitor the conformational changes which occur on a time scale of seconds or longer associated with crystal formation. This study was carried out over a period of nine months and reveals the importance of temperature in defining the rates and type of transformation which can occur on the form of the final crystalline state.

The first measurements of the rheological changes which occur during a photo cured epoxy acrylate/ acrylate mixture are reported. The profile of the viscosity with time is markedly different from that of a simple thermal cure and can be interpreted in terms of phase separation occurring during the initial course of the reaction. The gel time-intensity dependence follows a power law of the type $t_p \propto I^{-1.6}$ rather than the usual 1/2 power law. Evidence for the phase separated nature of the final material is also presented.

2 INTRODUCTION

Gelation and cure both involve the generation of supra-molecular structures either in solution or the solid state. The physical properties of the material will depend on the glass transition temperature $\{T_g\}$ of the polymer in that state and in the case of the cure of an epoxy resin, further reaction will convert the gel state to a glassy solid (vitrification). Photophysical measurements have been used extensively to probe the local mobility of polymer chains both in solution and the solid state, but are not usually sensitive to the longer range processes associated with gel and network formation[1-6].

Solutions of isotactic polystyrene [iPs] in benzyl alcohol [BA] quenched from 443 to 273K leads to the formation of a gel[7-9]. X-ray diffraction, infrared spectroscopy and electron and optical microscopy measurements have been performed on the gel and indicate that the polymer chains exist in an extended conformation[10-13]. The gel formed by cooling a solution of iPs in BA from a temperature of 403 - 343K at a maximum rate of 25K/min., leads to an unstable gel in which the chains adopt a 3-fold helical structure. A slow transformation is observed to occur to the more stable crystallizable lamellar form over a period of time. The gel structure can readily be converted into the closed helical form by heating to between 383 and 393K; the rate at which the process occurs depends critically upon the solvent and precise conditions used. The mechanisms whereby the interconversion process occurs are not fully understood and may either involve direct interconversion or, alternatively, a disruption of the extended helical structure prior to the formation of the closed helical form.

Conformational change in polystyrene has been studied by a wide range of techniques and measurements of the oligomers 2,4-diphenylpentane and 2,4,6-triphenylheptane have shown that in the isotactic isomer the tg and gt structures are more stable than the tt conformation, which has an energy 8kJ/mol higher than the former and accounts for less than 5% of the structures present[14-17]. Barriers for conformational change between 9 and 12 kJ/mol are observed for the oligomers. The most probable conformation in isotactic polystyrene has the tt and tg$^+$ or g$^-$t structures and leads to the generation of a helical form that favours the extended conformation. Ultrasonic measurements on solutions of atactic polystyrene indicate that a helix - coil transition can be induced on heating above approximately 340K, in agreement with earlier light scattering measurements.

The photophysical properties of polystyrene are dominated both in solution and in the solid state by inter- and intramolecular excimer formation[18-22]. The latter is associated with the occurrence of a specific conformation of the polymer chain in which two neighbouring phenyl groups share the excited state energy. Excimer formation has been investigated by a number of researchers interested in the configuration distribution, dynamics and photodegradation of polystyrene and can be explained in terms of a combination of local rotational isomeric and longer range diffusional motions of the chain. In the present study the iPs gel state has been investigated extensively in BA, which is not at first sight an ideal media for photophysical studies; however, it has the advantage of being a good solvent for the gel formation and is transparent in the infrared regions where the characteristic absorptions of the helical conformations occur. Moreover it has been found that BA plays an important role in determining the conformational structure of the gels and hence was selected for these investigations.

Photocuring is potentially an important environmentally friendly method of film formation, in that it converts monomer completely to solids avoiding the problems of solvent emission currently encountered with many solvent based systems. Photocuring has been used for many years and the technology and basic science associated with the processes are well developed. Cure monitoring or Smart processing has been used recently to optimize the conditions used for the generation of thermoset and more recently thermoplastic composite structures; dielectric measurements being used to probe the changes in mobility of the resin and relate this to corresponding variation of the macroscopic viscosity[23-28]. In this paper we report the use of a novel curometer to follow the changes which occur in the viscosity of the resin during the process of photocuring. Analysis of this data reveals the complexity of the reactions which are occurring and raises questions about the nature of the photo initiation process. The system selected for this study is based on a mixture of epoxy diacrylate with an acrylate; the difunctional monomer will act as a crosslinking agent whereas the concentration of the monofunctional monomer will, in principle, dictate the molar mass between crosslinks and hence the topography of the network formed[27-30]. Variation of the concentration of the photoinitiator and the intensity of the light determines the number of reaction sites generated and hence the speed of the polymerization.

3 PHOTOPHYSICAL STUDIES OF ISOTACTIC POLYSTYRENE

Materials

Isotactic polystyrene [iPs] of molar mass 9×10^5 (M_w) and $M_w/M_n = 5.2$ was obtained from Shell Chemicals. The polymer was purified by refluxing in decane for 3 hours under a blanket of nitrogen, then rinsed with methanol and dried in vacuo for 10 h. This procedure removed low molecular mass fractions stabilizer and antioxidants that might interfere with interpretation of fluorescence measurements. Absorption spectra were used to monitor the efficiency of this process. The tacticity was investigated by using a Bruker 270 MHz NMR and the ^{1}H spectra indicated approximately 95% of the polymer has the isotactic structure. Gel liquid chromatographic grade benzyl alcohol obtained from Aldrich Chemical Co. was used as the solvent; the purity was checked by using absorption spectra.

Preparation of Solutions and Gels

Benzyl alcohol containing 5,10 and 16% of iPs was heated to about 440K under a blanket of nitrogen to exclude oxygen, and an optically clear colourless solution was produced. Gel films were obtained by quenching a drop of the solution on precooled quartz plates held at 273K. Clear films of 40μm thickness, dry to touch, were obtained after equilibration for about 2 days at 273K. This procedure was rigorously followed to ensure reproducibility of the data.

Steady State Fluorescence Measurements

Fluorescence emission and excitation spectra of gels mounted on quartz plates were recorded using a Shimadzu RF-540 spectra fluorometer. In all of the films studied, the emission spectra were invariant with excitation wavelength. An excitation wavelength of 257 nm and a slit width of 5 nm were used in all the studies.

Fluorescence Decay and Time Resolved Measurements

Fluorescence decay and time resolved measurements were performed using a single photon timing technique. A slitwidth of 10 nm for the excitation and analysis beams was increased to 20 nm for the time resolved excitation experiments. The problem of scattered light was minimized by the introduction of a polarizer into the emission beam, and the films were mounted at 45° to the excitation beam. Measurements were performed at room temperature (298K). Approximately 10^4 counts were required in order to obtain accurate lifetimes and intensities by reconvolution analysis and include data from raising edge of the peak. Values of I(t) have been synthesized using a reconvolution function of the type:

$$I(t) = B_1 \exp(-t/\tau_1) + B_2 \exp(-t/\tau_2)$$

where τ_1, τ_2 are the lifetime parameters and B_1, B_2 are the pre-exponential functions. Data analysis was performed using the IBH software package which incorporates a non linear least squares fitting procedure and runs on a PDP 11/73 microcomputer. The values of χ^2 were used as the test of the goodness of fit and values of 1.2 or lower were interpreted as indicating the appropriateness of the kinetic model[31,32].

Fourier Transform Infrared Measurements

A Nicolet FFT 60 spectrometer fitted with a thermostated cell allowed investigation of the temperature and time dependence of the conformational structure of the iPs gel and involved subtraction of solvent from solution spectra[33].

Results and Discussion

Photophysics of Benzyl Alcohol. Benzyl alcohol is capable of absorbing light between 210 and 240 nm and fluorescing between 265 and 310 nm and, therefore, has the potential of transferring energy to polystyrene. Measurements of the fluorescence lifetime performed on BA and a solution in hexane indicate a single exponential decay invariant with wavelength. However, there has to be recognized the possibility of a small amount of solvent emission at 350 nm could interfere with the fluorescence of the iPs gel. A solution of atactic polystyrene was investigated to explore the possibility of interference effects and lifetimes were indentical with those published previously for solutions of non absorbing solvents. Diluting BA with hexane lowers the viscosity and consequently the monomer lifetime. The solution of atactic polystyrene exhibits a two component lifetime spectra, the shortest component being close to that for BA in hexane – it is important to differentiate between solvent and monomer emission. The second component has a lifetime comparable to that for excimer emission in polystyrene. The emission spectrum of the iPs varies reversibly with temperature in the region 273 - 306K, as shown in Figure [1]; however the excimer emission intensity is constant while the monomer emission intensity increases with increasing temperature. Reversibility of the spectra indicates that the system is in thermodynamic equilibrium. X-ray, infrared and optical studies of the morphology of the gel indicate that it is consistent with reversibility of the emission spectra[34–37].

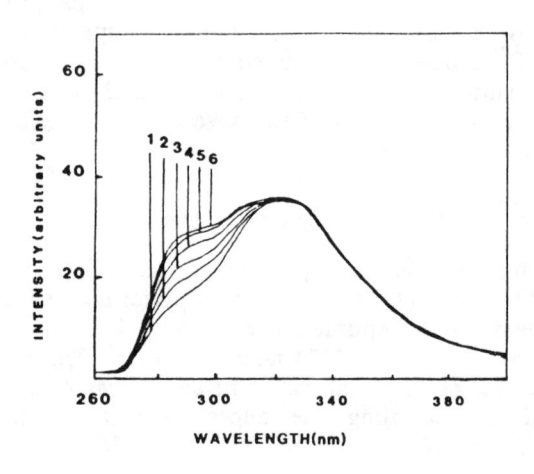

Figure (1) Fluorescence emision spectra of 5% w/w iPs/benzyl alcohol gel at 273, 277, 285, 289, 295 and 306K (1- 6) respectively.

Above 306K, the intensities of both the monomer (J_M = 285 nm) and excimer (J_E = 330 nm) emission are temperature dependent, as shown in Figure [2], and indicate that a non-radiative process is responsible for the changes in monomer emission intensity in the region 273 - 306K.

Above 306K, there is a marked decrease in the monomer emission intensity and a corresponding increase in the excimer emission intensity, as illustrated in Figure [3], and the gel becomes unstable and can change its viscosity and crystallize.

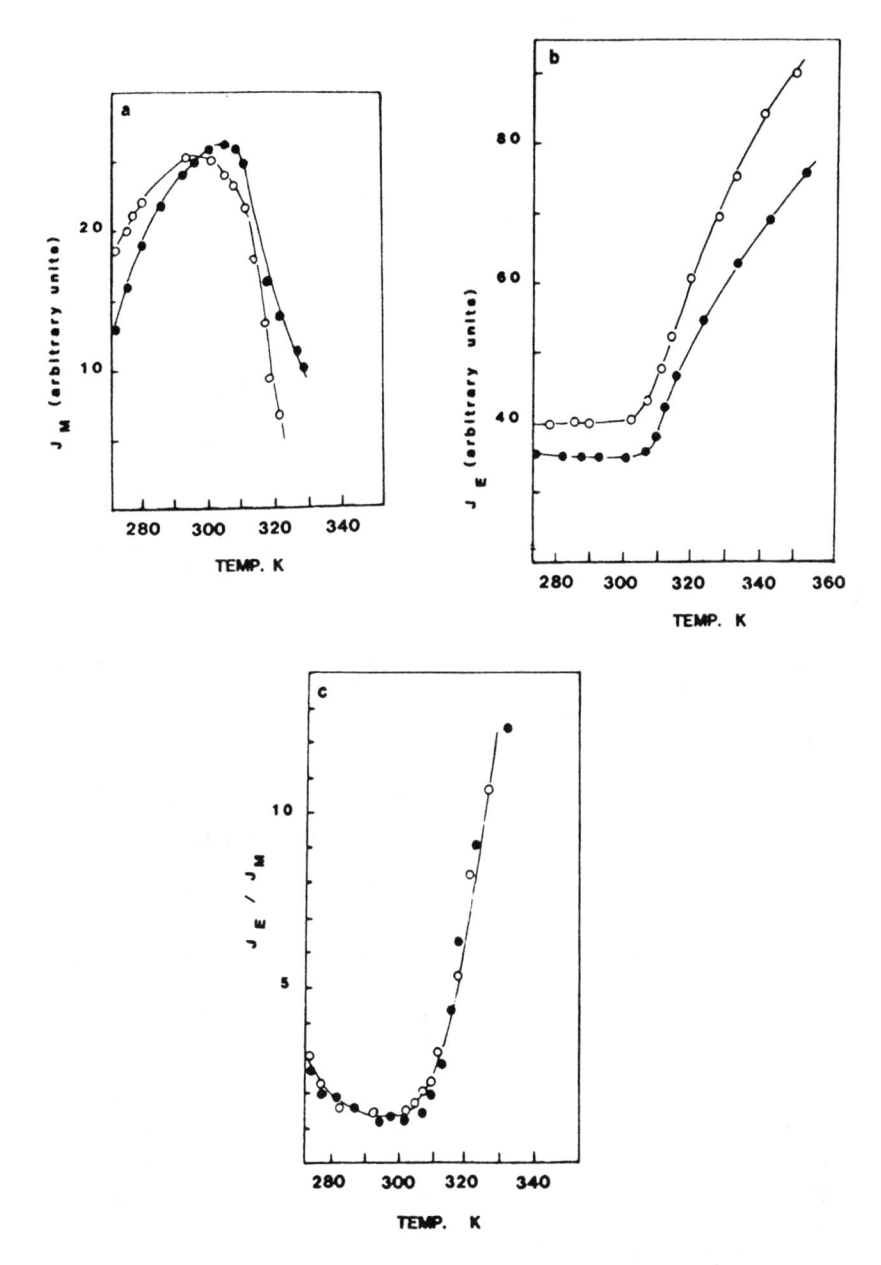

Figure (2) Dependence of monomer (a) and excimer emission (b) (intensity λ_{EM} = 285 and 330 nm) and the ratio (J_E/J_M) (c) on the temperature of a iPs gel of different concentrations; O-5%w/w, ●-10% w/w (iPs/benzyl alcohol).

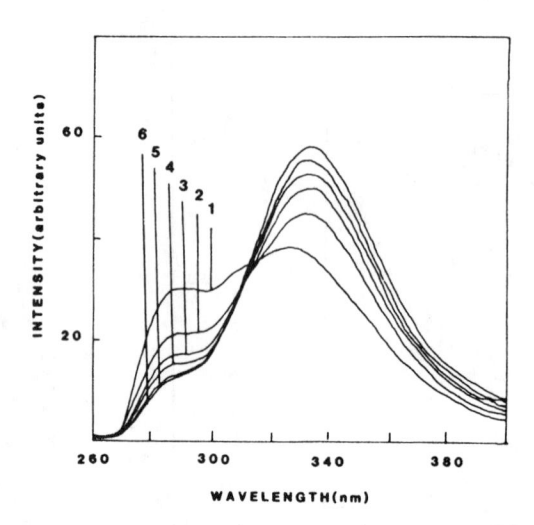

Figure (3) Changes of the fluorescence emission spectrum of a iPs (10% w/w) annealed at 318K: (1) 0min, (2) 4 Min, (3) 9 min, (4) 14 min, (5) 20 min, (6) 30 min.

Oxygen diffusion and quenching cannot explain the initial increase in the monomer emission and reversibility below 306K may be attributed to changes in the conformational distribution in the gel. Optical density, infrared, viscosity and NMR measurements indicate that above 313K the iPS chains undergo conformational changes that are time dependent and irreversible, supporting the assumption that the effects are directly related to changes in the conformational distribution. At low temperatures strong intramolecular interactions between neighbouring phenyl rings lead to a high probability of excimer formation. Benzyl alcohol at low temperatures is a poor solvent for iPS and hence phenyl-solvent interactions are unfavourable in comparison to intramolecular phenyl-phenyl interactions. Because of the irreversibility of the process above 313K, the measurements of the intensities were made with a heating rate of 0.4K/min and are believed to involve both changes in the local conformation and long range structure of the gel. The kinetic nature of the processes can be seen from changes in the spectra at 318K over a period of 30 minutes, Figure [3]. Variation in the rate of change, reflected in the ratio of the J_E/J_M as a function of time and temperature, Figure [4] indicates that the process cannot be represented by a single rate equation and implies a possible diffusional character to the process. A major part of the monomer emission will come from the solvent and hence reflects changes in the viscosity of the solvent. The fact that the ratio initially increases before decreasing indicates significant effects arising from polymer monomer emission. Above 318K, a combination of local conformational changes perturbed by the effects of intermolecular interactions between neighbouring polymer chains controls excimer formation. Below 318K, chain diffusion appears to be inhibited and conformational changes are reversible. Roots et al[38] have observed that the ratio of J_E/J_M varies with concentration and is the result of intermolecular excimer formation. Close correspondence between the variation of J_E/J_M with time and concentration implies that the dominant process is intra- molecular, the differences with concentration reflecting the intermolecular excimer. At 5% the polymer chains are just at the volume filling limit, whereas at 16% a stable gel will be formed. The variable helical structure of the gel makes it impossible to calculate the number of contacts; however, the formation of a gel at 5% indicates that at least two contacts must exist per chain.

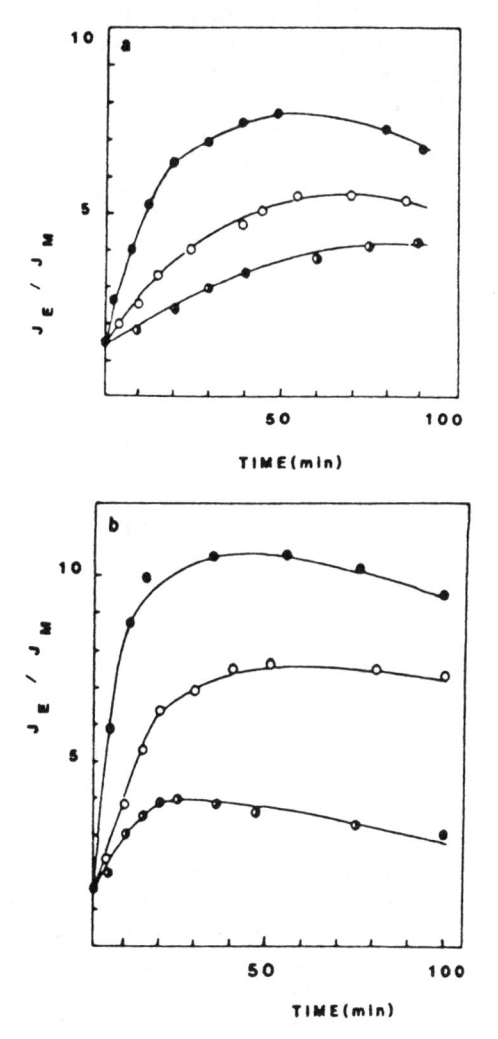

Figure (4) Kinetics of the ratio J_E/J_M during heating of a iPs gel (10% w/w) at (a) different temperatures, ●-318K 0-316K and ⊙-314K and at (b) different concentrations, ●- 16%, 0- 10% and ⊙- 5% w/w (iPs/benzyl alcohol), at 318K.

Annealing the iPs gel between 313 and 353K leads to a decrease in intensity of monomer emission which could be a consequence of enhanced oxygen quenching; however an increase in excimer emission intensity suggests that this is not the case. Increase in the excimer emission intensity occurs initially at 330 nm; however at longer times the intensity at higher wavelengths increases consistently with a shift of the maximum in the intensity, the rate at which the shift occurs increasing with increasing temperatures.

The gel obtained by quenching a solution from an initial temperature of between 383 and 393K produces iPs crystallizing in a 3 fold helix structure. A sample of iPS gel annealed at 383K for 3h. formed the 3 fold helix structure and has a characteristic excimer emission at 330 nm, shown in Figure [5b]. After annealing for 3 hours at 353K, two types of excimer can be distinguished Figure [5c], an excimer occurring at 330 nm associated with the 3 fold helix structure and a red shifted excimer with a maximum located at 345 nm ascribed to an extended structure.

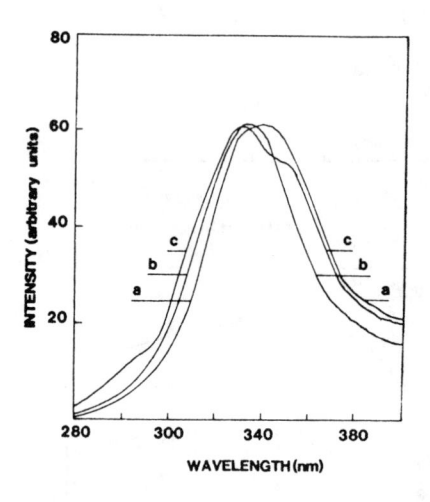

Figure (5) Fluorescence emission spectra of iPs gel (10% w/w) (λ_{EX} = 257 nm): after annealing at 343 K for 3 h., (b) after annealing at 383K for 3h., (c) after annealing for 3h at 353K.

The two excimer bands belong to two thermally selected conformations associated with different ordered chain structures. A detailed FTIR spectroscopic study of these solutions has indicated that the excimer at 330 nm can be associated with a 3 fold helix structure and the excimer at 345 nm to an extended structure in which solvent is incorporated. A more detailed discussion has been presented elsewhere[7].

<u>Fluorescence Decay Studies of iPs Gels.</u> Steady state fluorescence data on the gel can be divided into three regions: monomer emission intensity having a maximum at 285 nm and excimer emission intensities

at 330 nm, characteristic of the 3 fold helix form, and at 345 nm associated with the extended conformation, represented in Figure [6]. Fluorescence decay profiles obtained at different wavelengths corresponding to different characteristic conformations of the gel state of iPs in benzyl alcohol were measured. Fluorescence decay curves, Table (1), obtained from the initially formed gel in the region of the monomer and excimer emission intensity are similar to those for atactic polystyrene. Gupta et al[21] using methylene chloride as solvent found a monomer lifetime of 0.72 ns and an excimer lifetime of 12.5 ns with a value of 20.0 ± 2.5 ns in the solid state. In the gel state, the monomer emission at 285 nm is slower than that in polystyrene solution, however, the excimer lifetime is comparable to that for the solid, reflecting a slowing down of motion in the gel. Variation of the excimer lifetime with wavelength reflects a range of closely related excimer structures.

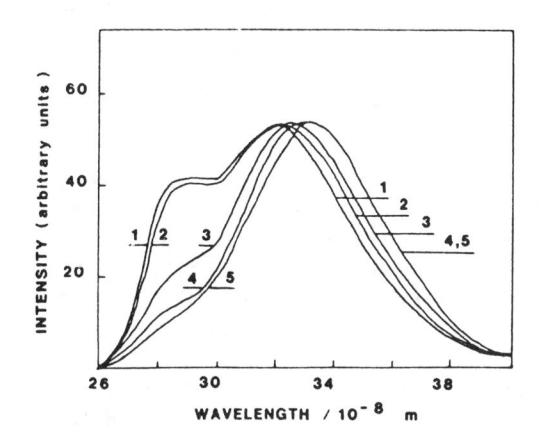

Figure (6) Steady state fluorescence spectra for iPs/benzyl alcohol after storage: (1) 1h, (2) 240h, (3)432h, (4)744h,(5) 864 h. Excitation at 257 nm, bandwidth 5 nm.

Annealing the gel at 318 and 345K increases the concentration of red-shifted excimer at 345 nm. Analysis of the fluorescence decay at 350 nm requires a third exponential with a lifetime of 30 ns, indicating the extended structure with incorporated solvent. The gel annealed at 383K exhibits a maximum in its emission intensity at 330 nm and can be adequately fitted to two components. The lifetimes indicate that the lower intensity component corresponds to the 3 fold helix form. Similar red shifts were observed in the excimer fluorescence of stretched films of atactic polystyrene and increase the lifetime of the long component from 21.0 ns in the unstretched to 22.6 ns in the stretched film.

Table 1   Fluorescence decay data for iPs/BA gel as a function of temperature below the sol-gel transition at emission wavelength 325 nm with excitation at 257 nm

Temp./K	τ_1/ns	τ_2/ns	B_1/%	B_2/%	χ^2
273	9.7±0.9	19.8±0.3	25.4	74.6	1.23
278	10.1±0.4	20.5±0.3	28.6	71.4	1.14
282	10.8±0.4	20.6±0.4	78.1	21.9	1.06
289	10.5±0.6	22.4±0.2	89.5	10.5	1.18
297	8.0±0.2	19.8±0.5	64.3	35.7	1.19
303	4.6±0.8	18.2±0.6	17.8	82.2	1.21

Long Time Relaxation of the Gel. In order to ensure that longtime effects observed over a period of up to 1656 hours are a consequence of the structure of the gel and not of changes in the solvent concentration, the samples were stored in an environment saturated with solvent. This ensures that any solvent lost from the gel is a consequence of shrinkage induced by conformational change rather than solvent evaporation. Initially the excimer is independent of wavelength, indicative of only one conformation being present. Systematic changes in the lifetime are observed, resulting from changes in the distribution of conformations contributing to the excimer spectrum at 330 nm. The increased lifetime correlates with the macroscopic observation of densification of the gel. After 432 hours storage, the emission spectrum exhibits the red shifted excimer reflected in an increase in the lifetime of the emission at 350 nm; approximately 30% of the excimers have changed into the long lived component, with a characteristic lifetime of 27 ns and reflect change of the extended conformation into the solvated 3- fold helix form.

Time Resolved Emission Spectra. Changes in the distribution of excimer states can be explored by using time resolved spectra. The red shifted excimer emission at 345 nm is clearly identified by using a time window of 79-152 ns which covers the whole of the emission spectra of the long lifetime component. The dotted curve in Figure [7] is the steady state fluorescence emission spectrum for the gel immediately after it was formed.

A very close correspondence of this line shape with that of the short lived component confirms that the extended form is the predominant species formed initially. The red shifted solvated excimer is only generated after a period of time. After a period of 744 hours a red shifted excimer with a maximum at 345 nm is observed using a time window of 96-198 ns. The excimer component at 325 nm is reduced in

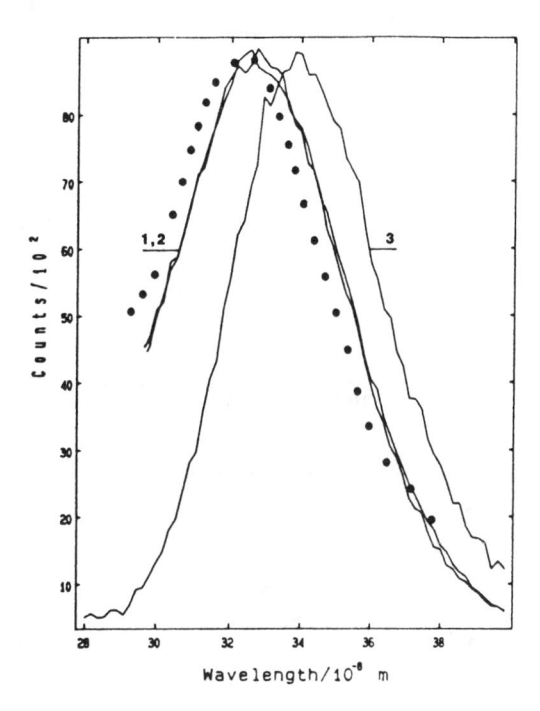

Figure (7) Time-resolved emission spectra for iPs/benzyl alcohol gel stored for 270 h; (1 and 2) t=0, time window 9ns, (3) t = 79ns, time window =73 ns. Dotted curve: steady state fluorescence emission spectrum for iPs/BA gel after preparation. Excitation is at 257 nm, bandwidth is 2nm.

intensity, being transformed into a new structure at 330 nm. After 1658 hours, the whole gel is transformed into the lowest energy structure and exhibits a maximum at 330 nm. The number of solvent molecules per monomer unit is calculated from densification data, as shown in Figure [8a], and this can be correlated with the change in lifetime of the component at 330 nm, shown in Figure [8b].

The decrease in the number of solvent molecules per monomer is consistent with the idea that the final form of the iPs has excluded solvent from the gel.

Phase Behaviour. Measurements of the phase structure of the iPs gels can be made using the scattering intensity of light. A break in the Arrhenius plot for the excimer fluorescence parameter, J_E/J_M, corresponding to a change of the slope from negative to positive was observed for the gel. An activation energy for the region above the break of 71 kJ/mol corresponds to a phase separation process. Sol-gel and gel-sol transition curves, cloud point temperatures, phase separation

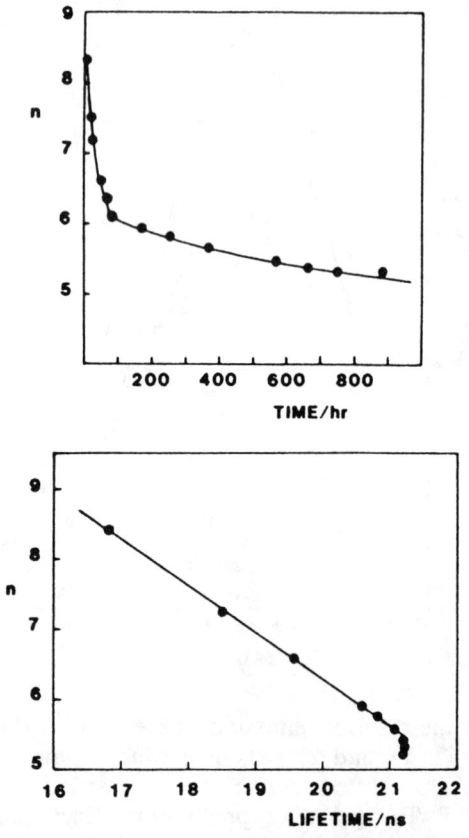

Figure (8) (a) Number of solvent molecules (n) per monomer unit in the chain of isotactic polystyrene as a function of time of storage of the iPs/benzyl alcohol gel. (b) Correlation of the lifetime of excimer fluorescence emision and the number of solvent molecules per monomer unit in the iPs chain.

diagrams and glass transition temperatures were measured. The enthalpy of gelation was calculated from the phase behaviour of the gel to be 83 kJ/mol per junction point.

Thermoreversible gelation of polymers follows the relationship between concentration and gelation temperature, firstly proposed by Eldridge and Ferry[39]:

$$\ln C = \frac{\Delta H}{RT} + \text{constant}$$

where C is the polymer concentration and the enthalpy change ΔH is the heat of gel formation per mole of junction points. A linear plot of log C as a function of the reciprocal temperature was obtained for gel formation. The enthalpy of gelation ΔH = 83 kJ/mol is similar to the value obtained for gelation of crystalline polymers. Similar values of the enthalpy were obtained from the gel-sol transition, 83 kJ/mol. The cloud point temperatures were obtained with a rate of heating of 5 deg/min. Two cloud point temperatures were observed separated by about 20 deg. The cloud point temperature depends on the rate of heating and only one value was observed for a heat rate of about 20 deg/min. The data points between the two cloud point diagrams are associated with the glass transition temperature, Tg, and agree well with the data for atactic polystyrene gels obtained by Hikmet et al[40]. This data confirms that there are phase changes occurring in the temperature range used in these studies. A very close correspondence is observed between the lower temperature transition; sol-gel transition and the reversible to irreversible transition in the gel state. It is therefore apparent that until the gel has passed through this phase transition, diffusional motion is severely inhibited and overall conformational changes of the chain leading to densification are inhibited. The picture which emerges indicates that the iPs gel below 310K is stable and redistribution of the structural forms occurs within chains that are effectively locked in space. Above 310K, the extended structure is generated and the chains undergo reptation motion. The processes in this region are irreversible, and the gel properties depend on the thermal history of the sample. The time resolved emission spectra reveal the existence of a third excimer peak, and the densification process of the gel involves the excimer transformation into a red shifted solvated exciplex, which itself transforms into the compact 3 fold helical conformation. As these processes occur, the matrix undergoes densification with the exclusion of solvent, which implies a change in the solvent sheath around the polymer. Adiabatic compressibility measurements on similar systems have shown that the solvent will interact with the polymer chain for a period of 10^{-5} s, which is sufficient for it to influence the lifetime of fluorescence and is also consistent with the NMR data.

Comparison of infrared, X-ray, and fluorescence data indicates that the red shifted exciplex structure is an extended helical structure in which solvent is incorporated. This study illustrates how fluorescence measurements can provide insight into the nature of the structures responsible for the formation of iPs gels and crystals.

4 CURE MONITORING OF PHOTOCROSSLINKING EPOXY-ACRYLATE RESINS

Materials

Actocryl-400, an aromatic epoxy diacrylate resin based on diglycidyl ether of bisphenol A(DGEBA) supplied by Ancomer, Manchester, UK[41], was used as the main matrix material. This resin can be crosslinked by photo as well as thermal initiation in combination with other vinyl monomers or low molecular weight acrylates. In the present case, n-butylmethacylate (nBA) was used as a comonomer and it was added in various proportions in the range 10 to 80% by weight. The photoinitiator was dimethoxy phenol acetophenone (Irgacure 651, Ciba Geigy) and its concentration was 5% by weight of the total resin matrix. The materials were degassed prior to being used in the experiments. The initiator was first dissolved in nBA to which the Actocryl resin was added and mixed in a dark room with low level of red illumination. The mixture was poured into a specially designed cell transparent to UV radiation and capable of being used for viscosity measurements. The cell consisted of thin microscope glass slides stuck together with a glass spacer (1.0 mm) using Araldite. The cell containing the photocurable formulation was clamped firmly by its sides to a base plate and the probe, a flattened stainless steel wire (0.6 mm thick), was allowed to freely vibrate in the liquid mixture. The cell was illuminated from all sides using an air cooled 200W medium pressure mercury lamp (Osram, Germany) provided with a water cooled neutral density filter and mounted on an optical bench. The area of illumination was more than twice the size of the cell. The details of the method used for viscosity/gel time measurements using this 'Strathclyde Rheometer' are reported elsewhere[42].

Results and Discussion

The viscosity of the photocuring mixture was monitored continuously during exposure, as shown in Figure [9], at six different intensities for a composition containing Actocryl : nBA ratio of 60:40 with 5% initiator. There are several interesting features to be noted from these measurements;

(a) there is an induction period before the start of the polymerization which depends on intensity. This effect may be ascribed to two effects: random initiation and oxygen scavenging. Oxygen scavenging will inhibit polymerization and hence delay the production of polymer. Random initiation will lead to production of a dilute solution of polymer at low light intensities. The viscosity will increase rapidly only when the concentration of linear polymer has exceeded a critical level; $C [M] \approx 4\text{-}6$, where C is the concentration in grams per decilitre and M is the molecular weight. Below this limit the polymer molecules will behave as an ideal solution, whilst above this limit corresponding to critical entanglement gelation will start to occur and a marked increase in viscosity is observed. Low levels of illumination will initially generate a small number of reactions and hence polymer molecules, further reaction leading to and increase viscosity.

(b) there is the appearance of a peak following the initial rapid rise in viscosity and this is especially marked at high intensity.

(c) there seems to be a threshold value of intensity necessary for the curing to proceed to completion.

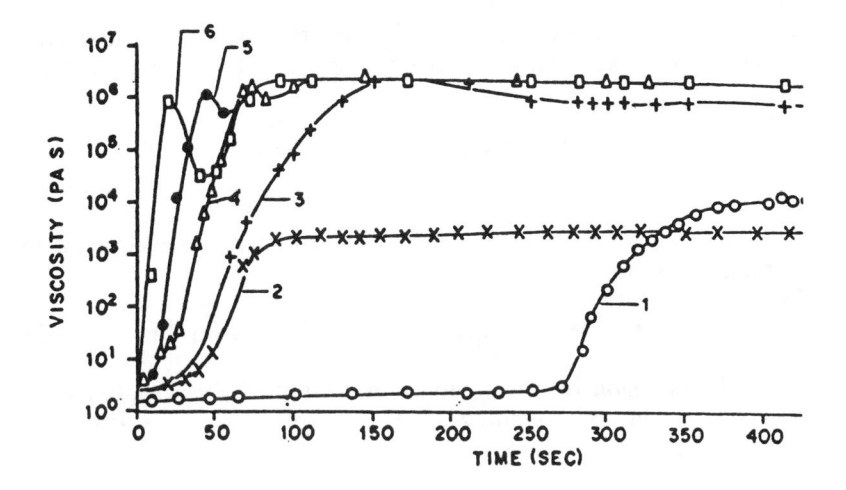

UV CURE MONITORING A400-nBA, 5% INITIATOR

Figure (9) Curing curves for the epoxy-acrylate resin (60/40), 5% initiator at various intensities; curves 1 to 6 correspond to increase of relative intensity from 1, 1.5, 2, 3, 4 and 5 respectively times the original dose (Temp = 303K).

In order to understand these results, it is necessary to first look into the mechanism and kinetics of the photopolymerization/crosslinking process. The various steps in photocuring are (i) absorption of light by photoinitiator, (ii) generation of free radicals, (iii) initiation of polymerization and oligomer formation, and (iv) formation of crosslinked product. Typically the photoinitiation rate (R_p) is given as[30,31]

$$R_p = \phi R_{absorp} = 2.303 \phi \epsilon [S] I_0 10^{-\epsilon l} \approx 2.303 \phi \epsilon [S] I_0 \qquad -(1)$$

for small thickness where ϕ is the quantum yield for the initiator, ϵ the absorption coefficient at the incident wavelength, [S] is its concentration, I_0 the incident light intensity and l is the thickness. However, absorbed oxygen inhibits the process of initiation by scavenging the free radicals generated. If [H] is the inhibitor concentration then we have

$$[H] = [H_0] - R_i t \qquad -(2)$$

Thus the onset of polymerization would take place at [H] ≈ 0 or from equations (1) and (2)

$$t_{onset} = [H_0] / \{2.303 \phi \epsilon [S] I_0\} \qquad -(3)$$

Now considering the usual free radical polymerization kinetics to operate after the initiation step one can derive the rate of polymerization as

$$-\frac{dM}{dt} = (k_p^2/k_t)^{1/2} R_i^{1/2} [M] = (k_p^2/k_t)^{1/2} \{2.303\ \phi_s\ \epsilon\ [S]\ I_0\}^{1/2} M \quad -(4)$$

where the various symbols have their usual meaning. The equation (4) on integration yields

$$\log(M_0/M_t) = (k_p^2/k_t)^{1/2} \{2.303\ \phi_s\ \epsilon\ S\ I_0\}^{1/2} t \quad -(5)$$

This brief discussion of the photopolymerization process suggests that the time to attain upper limiting value of viscosity should be dependent on intensity as

$$t_{peak} = A\ [I_0]^{1/2} \quad -(6)$$

It has been suggested that the above should be corrected for the onset time in order to observe $1/2$ power dependence[20]. In order to confirm the prediction, a plot of log t_{gel} vs log I as well as log ($t_{gel} - t_{onset}$) vs log I was made. In both cases the graph shows much stronger intensity dependence, $t_{peak} \propto I^{1.2}$ or $I^{-1.6}$, than expected from above equations. This suggests that the process of photocuring in this present case does not follow a simple mechanism and/or kinetics presented above should be modified to include the effects of random initiation. It is relevant to mention that even in the case of thermal curing of epoxy-acrylates we have observed that gel time also depends on the initiator concentration (peroxide for thermal case) as I^n where n = 1.6 to 1.8. These observations suggest that the curing reaction of epoxy-acrylate may involve, in addition, processes such as radical trapping, phase separation and/or spinoidal decomposition. The occurrence of a peak in the curing curve strongly suggests the possibility of phase separation[44], but could also be associated with the formation of a non-equilibrated gel material.

In order to investigate the possible phase morphology, these films were examined by optical polarizing microscopy and scanning electron microscopy. As such these photocured films are optically clear and transparent for all compositions suggesting that the phase domain size is quite small (of the order of 0.5μm or less). Also both the phases can have very similar refractive index values which are so close that there is little optical contrast. In order to increase the contrast these films were soaked in a dilute (0.2%) solution of Eosin - red dye in acetone for $1/2$ hr and then washed. Since the dye is expected to penetrate preferentially in acrylate rich domains this would increase the contrast. The micrograph of such a film provides clear evidence for a two phase morphology. This type of morphology was, however, only observed for samples with greater than 60% nBA. Further work is in progress on morphological investigations on photocrosslinked materials.

5 SUMMARY AND CONCLUSIONS

Cure monitoring of the UV cured epoxy-acrylate resins shows that the photo-crosslinking process involves the simultaneous occurrence of several events. Phase separation occurs for nBA rich systems. The type of morphology and network formed using high intensity illumination and short exposure time is quite different from that of low intensity and long exposure time and such differences can be explained as a failure of the reciprocity law and the occurrence of phase separation.

6 ACKNOWLEDGEMENTS

One of us (BW) wishes to thank the Royal Society and the University of Strathclyde for financial support for this study. One of us (SR) wishes to thank the British Council for support.

7 REFERENCES

1. I. Soutar, Polymer International, 1991, 26, 35.
2. H. Itagaki, Macromolecules, 1991, 24, 653.
3. J.M. Torkelson, M. Tirrel and C.W. Frank, Macromolecules, 1984, 17, 1505.
4. W.C. Tao and C.W. Frank, Macromolecules, 1990, 23, 3275.
5. S.N. Semerak and C.W. Frank, Macromolecules, 1984, 17, 1148.
6. S. Mani, M.F. Malone, H.M. Winter, J.L. Halary and L. Monnerie, Macromolecules, 1991, 24, 5451.
7. B. Wandelt, D.J.S. Birch, R.E. Imhof, A.S. Holmes and R.A. Pethrick, Macromolecules, 1991, 24, 5141.
8. B. Wandelt, Polymer, 1991, 32, 2707.
9. B. Wandelt, D.J.S. Birch, R.E. Imhof and R.A. Pethrick, Polymer, in press, 1993.
10. J.H. Aubert, Macromolecules, 1988, 21, 3468.
11. H.M. Tan, B.H. Chang, E. Baer and A. Hiltner, Eur. Polym. J., 1983, 19, 1021.
12. M. Girolamo, A. Keller, K. Miyasaka and N. Overbergh, J. Polym. Sci. Polym. Phys. Ed., 1976, 14, 39.
13. P.R. Sundararajan, Macromolecules, 1979, 12, 575.
14. R.A. Pethrick, Prog. Polym. Sci., 1983, 9, 197.
15. F.C. De Schryver, L. Moens, M. Van der Auweraeer, N. Boens, L. Monnerie and L. Bokobza, Macromolecules, 1982, 15, 64.
16. B. Froelich, C. Noel, B. Yasse and L. Monnerie, J. Chem. Phys. Lett., 1976, 44, 159.
17. B. Froelich, C. Noel, B. Yasse and L. Monnerie, J. Chem. Soc. Faraday Trans. 2, 1978, 74, 445.

18. W. Klopffer, 'Photophysics of Aromatic Molecules', J.B. Birks, Ed., Wiley-Interscience, New York, 1970. J.E. Guillet, Polymer Photophysics and Photochemistry', Cambridge University Press, Cambridge, 1985. Ch.L. Hoyle, J.M. Torkelson, Eds., 'Photophysics of Polymers', ACS Symposium Series 358, American Chemical Society, Washington, DC, 1987.
19. I. Soutar, D. Phillips, A.J. Roberts and G. Rumbles, J. Polym. Sci. Polym. Phys. Ed., 1982, 20, 1759.
20. D. Phillips, A.J. Roberts and I. Soutar, Macromolecules, 1983, 16, 1593.
21. M.C. Gupta, A. Gupta, Y. Horwitz and D. Kliger, Macromolecules, 1982, 15, 1372.
22. M. Gupta and A. Gupta, Polym. Photochem., 1983, 3, 211.
23. S.D. Senturia and N.F. Sheppard. Adv. Polym. Sci., 1986, 80, 3.
24. "Flow and Cure of Polymers - Measurement and Control". RAPRA Technology Ltd. Reports, UK, 1990.
25. C. Carlini, P.A. Rolla and E. Tombari, J. Appl. Polym. Sci., 1990, 41, 805.
26. G. Piews and R. Phillips, J. Coat. Technol., 1979, 59, 69.
27. D.R. Pemberton and A.F. Johnson, Polymer, 1984, 25, 529.
28. J. Guillet, 'Polymer Photophysics & Photochemistry', Cambridge University Press, Cambridge 1985, p.87, 81.
29. S.P. Pappas, 'U.V. Curing - Science and Technology', Vols. I and II, Techno. Market Corpn., Norwalk, Connecticut, USA, 1980.
30. C.G. Roffey, 'Photopolymerization of Surface Coatings', John Wiley & Sons, Chichester, UK, 1982.
31. D.J.S. Birch, R.E. Imhof and C.J. Guo, Photochem. Photobiol. A. Chem., 1988, 42 223.
32. D.J.S. Birch and R.E. Imhof, Annal. Instrum., 1985, 14, 293.
33. P.C. Painter, R.E. Kessler and R.W. Snyder, J. Polym. Sci. Polym. Phys. Ed., 1980, 18, 723.
34. E.D.T. Atkins, M. Hill, D.A. Jarvis, A. Keller, E. Sarhene, J.C. Shapiro, Colloid Polym. Sci., 1984, 22, 262.
35. P.R. Sundararajan, Macromolecules, 1979, 12, 575.
36. P.R. Sundararajan, N.J. Tyrer and T.L. Bluhm, Macromolecules, 1982, 15, 286.
37. P.R. Sundararjan and N.J. Tyrer, Macromolecules, 1982, 15, 1004.
38. J. Roots and B. Nystron, Eur. Polym. J., 1979, 15, 1127.
39. J.E. Eldridge and J.D. Ferry, J. Phys. Chem., 1954, 58, 992.
40. R.M. Hikmet, S. Callister S. and A. Keller, Polymer, 1988, 29, 1378.
41. Ancomer Technical Bull. A-400, Ancomer, Clayton, Manchester, UK.
42. S. Affrossman, A. Collins, D. Hayward, E. Trottier and R.A. Pethrick, Transaction JOCCA, 1989, p.452.
43. W.D. Cook, J. Appl. Polym. Sci., 1991, 42, 2209.
44. A. Ya. Malkin and S.G. Kulichikhin, Adv. Polym. Sci., 1991, 101, 218.

Luminescence in Synthetic Polymers

D. Phillips

DEPARTMENT OF CHEMISTRY, IMPERIAL COLLEGE, LONDON SW7 2AY, UK

SUMMARY

This paper presents a survey of the fluorescence of synthetic polymers and applications, with examples drawn from the author's own laboratory. Fluorescence spectral, intensity, decay and anisotropy measurements provide a means of studying order, molecular motion, energy transfer and migration, and heterogeneity.

INTRODUCTION

Many simple synthetic polymers such as poly(ethylene) and poly(propylene) in a pure state will exhibit only σ-σ^* absorptions in the high-energy UV region, where most organic molecules will absorb. Such excitations in general lead to photochemical reaction rather than luminescence, and excited states will thus be very short. Here we focus attention on species that absorb in the spectral region from, rather arbitrarily, say 250 nm to longer wavelengths, where luminescence may be an additional fate of photoexcited species, which will in general be longer lived.

A convenient classification of chromophores has been suggested by Somersall and Guillet[1] in which distinction is made between polymers in which the repeat unit contains an absorbing (and emitting) unit, termed type B polymers, and those termed type A in which isolated chromophores are attached to a polymer chain as an end group or minor component of a copolymer. As a typical example of a type B polymer, poly(styrene) may be cited, one member of a general class in which pendant aromatic groups are arranged along the backbone in a random (atactic), alternating (syndiotactic), or regular (isotactic) fashion.

The isolated chromophores in a polymer type A are in general adventitious, being due to end groups resulting from the polymerization process; commercial additives such as plasticizers, antioxidants, and pigments; products of thermal oxidation; or occasionally probes added deliberately. Typical examples of type A chromophores are the ketonic species present in thermally oxidized poly(propylene).

Attention is here confined to type B polymers.

In type B polymers the structural constraints of the polymer chain tend to confine the chromophores in spatial positions such that they can be expected to exhibit strong mutual interactions. These may depend strongly upon the relative orientation of the interacting chromophores, and the orientations themselves will usually be dependent upon the conformation of the polymer chain. Interaction between the excited state chromophore and a neighbouring ground state can give rise to excimer (excited dimer) formation, which proves to be a powerful diagnostic of interacting molecules.[2]

The salient features of excimer formation are represented in Fig. 1. Aromatic molecules at large separations, that is at separations much greater than 4Å, may be considered as isolated entities. Consequently, if the aromatic molecules are in an excited state the fluorescence is unaffected by the presence of other molecules. For small separations, less than 4Å, repulsive potentials $(R(r))$ and $(R'(r))$ will exist between molecules in their ground state and between molecules in the ground and excited states. The existence of these repulsive potentials, in general, prevents the formation of complexes. However, for the interaction between ground and excited state molecules an attractive potential $(V(r))$ may be obtained due to configuration interaction between resonance and exciton-resonance states. The combination of repulsive and attractive potentials may form the excimer state shown by the potential well in Fig. 1. The fluorescence from the 'excimer' state will thus be unstructured (since a corresponding ground state complex does not exist) and at a lower energy than the corresponding monomer emission. In general, excimer formation can occur whenever aromatic chromophores adopt a face-to-face coplanar arrangement with a separation of 0.3-0.35 nm.

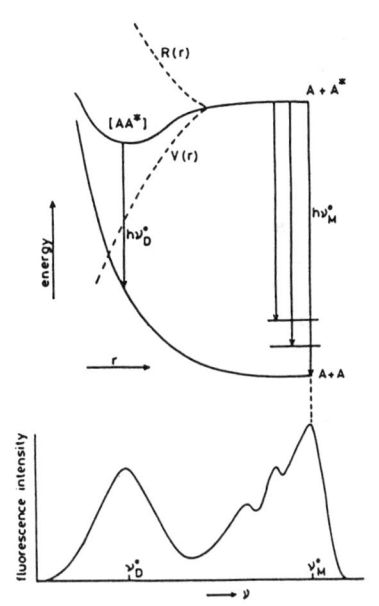

Fig. 1. Energy diagram for excimer formation.

The excited state properties, principally fluorescence, of probe molecules can be used to great effect to study molecular motion, order, and energy migration in synthetic polymer systems. Thus, as an example, measurement of the relative intensities of fluorescence of a probe molecule polarized parallel to and perpendicular to the plane of linearly polarized exciting radiation as a function of orientation of a solid sample yields information concerning the ordering of polymer chains. In a mobile system, similar time-resolved polarization studies yield information on the rotational relaxation of the probe molecule. The kinetics of formation and decay of excimers in vinyl aromatic polymers can yield information on energy migration and segmental motion, although the interpretation can be difficult (see below).

EXPERIMENTAL

Most of the experiments described here utilize fluorescence measurements, time-correlated single photon counting being the technique of choice.[3] The method relies on the basic concept that the probability distribution for emission of a single photon following excitation gives the actual intensity against time distribution of all photons emitted; thus by sampling the time of single-photon emission following a large number of excitation pulses, the probability distribution is created.

The experiment is carried out as follows, with reference to Fig. 2. A trigger which could be a photomultiplier, an antenna pick-up, or a logical synchronizing pulse from the electronics pulsing the laser excitation source, generates an electrical pulse at a time exactly correlated with the time of generation of the optical pulse. The trigger pulse is routed, usually through a discriminator (CFTD) to start input of the time-to-amplitude converter (TAC), which initiates charging of a capacitor. In the meantime the laser pulse excites the sample, which subsequently fluoresces. An aperture is adjusted so that at most one photon is 'detected' for each exciting event. The signal resulting from this photon stops the charging ramp in the TAC, which puts out a pulse, the amplitude of which is proportional to the charge in the capacitor, and hence to the time difference between START and STOP pulses. The TAC output pulse is given a numerical value in the analogue-to-digital converter and a count is stored in the data storage device in an address corresponding to that number. Excitation and data storage are repeated in this way until the histogram of number of counts against address number in the storage device has enough data so that it represents, to some required precision, the decay curve of the sample. If deconvolution is necessary, the time profile of the excitation pulse is collected in the same way by replacing the sample by a light scatterer. It is usually most convenient to run the experiment in reverse, so that the emitted photon starts the voltage ramp in the TAC and the next laser pulse stops the ramp.

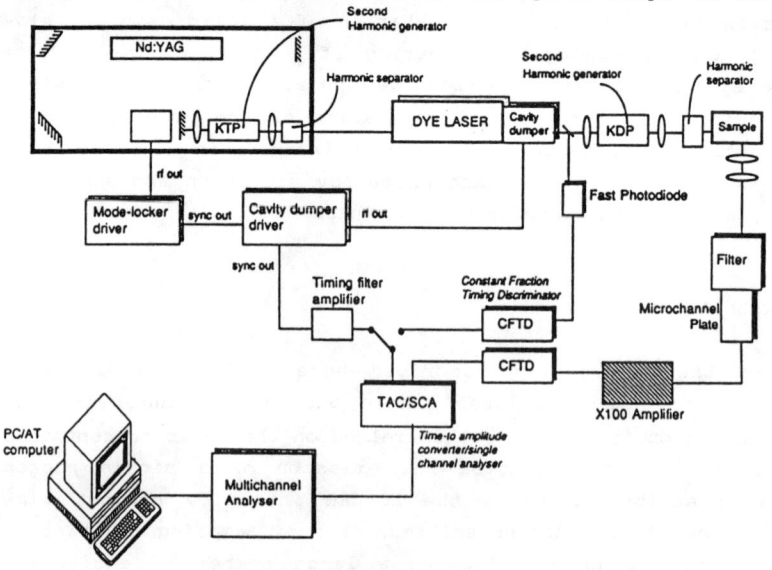

FIGURE 2 Time-correlated single-photon counting spectrometer based on a CW mode-locked Nd:YAG laser.

APPLICATION OF MEASUREMENTS

1. Order: (p-phenylene diacrylic acid).

The polyester PPDA below is a photoresist which on exposure to UV forms solid films by cycloaddition to give predominantly the β-truxinate (head-to-head) dimer. The predominance of this isomer and the very high photochemical yields in the system indicated a considerable degree of chromophore correlation. Since the films are amorphous, such correlations can extend only over pairs of small clusters of PDA groups: they nevertheless determine the photochemical behaviour of the material. This finding is not restricted to PPDA; the behaviour of other photopolymers also implies a degree of chromophore association in the solid. An attempt was thus made to monitor the process by which the state of chromophore aggregation in the matrix is produced.[4] In PPDA, the PDA chromophore fluoresces and is known to give excimer-like emission in solution.

$$\left[\begin{array}{c} -OOCCH=CH-\langle O \rangle- \\ -CH=CHCOOC_2H_4O-\langle H \rangle-OC_2H_4- \end{array} \right]$$

PPDA

This was shown by time-resolved measurements to be true dimer emission, indicating that the sites were preformed.[4] This observation permitted a calculation of the fraction of aromatic chromophores in solution of the polymer which existed in the associated form, the result for concentrations below 5×10^{-4}M being 0.26. This means that in the isolated PPDA molecule on average every fourth chromophore takes part in an intramolecular contact, and it implies a much higher contact frequency than would be expected in a chain of non-interactive segments. Support for this conclusion was found in the combined results of gel permeation chromatography and light scattering. Elution of PPDA in a g.p.c. column calibrated with narrow polystyrene fractions indicates for PPDA a polystyrene-equivalent weight average molecular weight of $M_W = 33000$ and a radius of gyration of $R_G =$

330Å. Light scattering in dichloroethane led to a similar radius of gyration (R_G = 350Å), but to a molecular weight of M_W = 155000. With this molecular weight PPDA occupies the same molar volume in solution as a polystyrene molecule of molecular weight M_W = 33000. The PPDA coil is, therefore, about five times as compact as that of polystyrene.

2. **Molecular motion: copolymers of acrylates and vinyl naphthalenes**.

The fluorescence depolarization method has been successfully applied to the investigation of relaxation processes of polymers. In order to study such processes a chromophore must be covalently bound to the polymer; by labelling specific sites of the polymer specific relaxation processes may be examined. However, there are a number of difficulties associated with this procedure.

a) the size of the chromophore should not impede the rotational motion of the polymer segment under consideration.
b) the label may not reflect the motion of the polymer segment if it is capable of independent motion.
c) it is not possible to assume that the fluorescence lifetime and the initial anisotropy of the chromophore when bound to the polymer is the same as that of the unbound chromophore.
d) the excited state lifetime of the chromophore must be of a comparable magnitude to the relaxation time of the process to be studied.

The problems outlined above can be readily overcome by judicious choice of chromophore.

The fluorescence decay times of excited states are such that the fluorescence depolarization technique may only be used to examine relatively high frequency relaxation processes of polymers. Consequently the fluorescence depolarization has been primarily limited to the study of relaxation processes of polymers in solution. The anisotropy of a system, $r(t)$, is derived from measurements of the fluorescence decays with polarizations parallel and perpendicular to the polarization of excitation:

$$r(t) = [I_{||}(t) - I_{\perp}(t)]/[I_{||}(t) + 2I_{\perp}(t)] = D(t)/S(t)$$

Time-resolved fluorescence anisotropy measurements can provide detailed information on the reorientation dynamics of molecules in solution. Until recently, however, this information has been limited to single rotational correlation times which are only strictly appropriate for the diffusion of spherically symmetric molecules. Improvements in instrumentation and data analysis techniques during the last decade have led to increasingly accurate measurements of fluorescence lifetimes. These capabilities also have led to parallel improvements in determinations of fluorescence anisotropies.

The advances in time-resolved techniques have fostered a reexamination of theories of the rotational motions of molecules in liquids. Models considered include the anisotropic motion of unsymmetrical fluorophores, the internal motions of probes relative to the overall movement with respect to their surroundings, the restricted motion of molecules within membranes (e.g. wobbling within a cone), and the segmental motion of synthetic macromolecules.[5] Analyses of these models points to experimental situations in which the anisotropy can show both multiexponential and nonexponential decay. Current experimental techniques are capable in principle of distinguishing between these different models. It should be emphasized, however, that to extract a single average rotational correlation time demands the same precision of data and analysis as fluorescence decay experiments which exhibit dual exponential decays. Multiple or nonexponential anisotropy experiments are thus near the limits of present capabilities and generally demand favourable combinations of fluorescence and rotational diffusion times.[6]

An example is cited below of a study on the copolymers

a)   methyl methacrylate/acenaphthylene (PMMA/ACE)
b)   methyl methacrylate/1-vinyl naphthalene (PMMA/1-VN)
c)   methyl acrylate/acenaphthylene (PMA/ACE)
d)   methyl acrylate/1-vinyl naphthalene (PMA/1-VN)

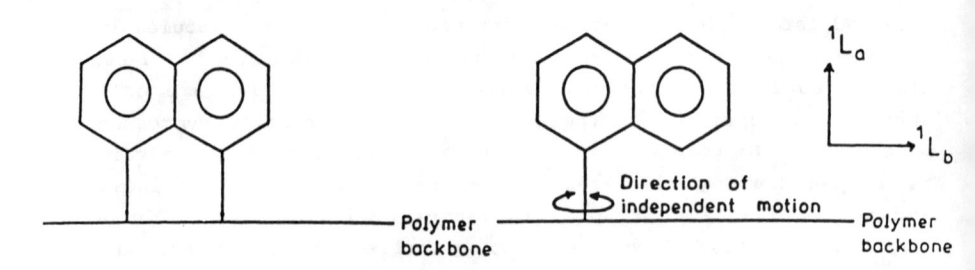

Fig. 4 Alignment of the transition dipoles and the direction of the independent motion of the 1-vinyl naphthalene chromophore relative to a polymer backbone

Results are summarized in Table 1.

TABLE 1

	T/K	τ_F/ns	r_0	τ_R
PMA/ACE	298 ± 2	17.4 ± 0.2	0.10 ± 0.01	0.8 ± 0.3
	260 ± 2	17.4 ± 0.3	0.10 ± 0.01	1.3 ± 0.2
	245 ± 2	17.4 ± 0.3	0.11 ± 0.01	1.8 ± 0.3
	230 ± 2	17.5 ± 0.3	0.12 ± 0.02	2.5 ± 0.3
PMA/1-VN	298 ± 2	15.1 ± 0.1	0.13 ± 0.01	0.5 ± 0.1
	275 ± 2	14.9 ± 0.1	0.13 ± 0.01	0.8 ± 0.1
	260 ± 2	14.8 ± 0.1	0.14 ± 0.01	1.0 ± 0.2
	245 ± 2	14.9 ± 0.1	0.14 ± 0.01	1.3 ± 0.3
	230 ± 2	14.9 ± 0.1	0.15 ± 0.01	1.7 ± 0.3
PMMA/ACE	298 ± 2	15.5 ± 0.1	0.13 ± 0.01	1.3 ± 0.1
	275 ± 2	15.7 ± 0.1	0.13 ± 0.01	2.2 ± 0.2
	260 ± 2	15.4 ± 0.2	0.13 ± 0.01	3.2 ± 0.5
	245 ± 2	15.5 ± 0.2	0.13 ± 0.01	4.5 ± 0.7
	230 ± 2	15.6 ± 0.1	0.11 ± 0.02	5.6 ± 0.7
PMMA/1-VN	298 ± 2	15.9 ± 0.2	0.15 ± 0.01	1.3 ± 0.2
	275 ± 2	15.6 ± 0.2	0.16 ± 0.01	2.2 ± 0.5
	260 ± 2	15.5 ± 0.1	0.14 ± 0.01	2.7 ± 0.3
	245 ± 2	15.4 ± 0.1	0.15 ± 0.01	3.6 ± 0.5
	230 ± 2	15.4 ± 0.1	0.16 ± 0.01	4.9 ± 0.7

Averaging all the determinations for the initial anisotropy for each polymer sample leads to the following values for excitation at 300 nm:

PMMA/ACE $r_o = 0.13 \pm 0.01$ PMA/ACE $r_o = 0.11 \pm 0.01$

PMMA/1-VN $r_o = 0.15 \pm 0.01$ PMA/1-VN $r_o = 0.14 \pm 0.02$

These are in excellent agreement with values obtained for polymers with similar compositions. Initial anisotropies are expected to have the value of 0.4. However, the first and second excited states of naphthalene and its derivatives are, in the Platt notation, designated ^1L_b and ^1L_a respectively. The transition dipole moments for absorption into these bands are directed along the long (^1L_b) and short (^1L_a) axes of the aromatic rings (Figure 4). Irradiation at 300 nm produces excitation of both absorption bands and so naphthalene and its derivatives, when excited at this wavelengths, can be considered to have a planar rather than a linear absorption oscillator.

The 1-vinyl naphthalene chromophore, unlike the acenaphthalene chromophore, would appear to be capable of motion independent of the polymer backbone. It is at first surprising that the anisotropy decay for the 1-vinyl naphthalene labelled polymer was shown to have the same form and value of rotational relaxation constant as those of the acenaphthylene labelled polymers. However as stated above the fluorescence polarization properties of the 1-vinyl naphthalene chromophore are explained in terms of two emitting transition dipoles. These transition dipoles are aligned parallel and perpendicular to the bond about which the chromophore undergoes independent motion of the polymer backbone (see Figure 4). Rotation about this bond cannot lead to these two transition dipoles interchanging their directions relative to the polymer backbone. Consequently for the 1-vinyl naphthalene labelled polymers, as with the acenaphthylene labelled polymers, it is only segmental motions which lead to depolarization.

For the poly(methacrylates) and poly(acrylates) the α and β relaxations are associated with segmental motions of the polymer and independent motions of the ester substituents respectively. The merging of these transitions at high frequencies or temperatures corresponds, at the molecular level, to the incidence of co-operative motion between the substituent and polymer backbone. Consequently, it is to be expected that in solution, the high frequency motions of both polymer chain and fluorescent label will assume a co-operative form characterized by a single relaxation process/time.

The activation energies derived from the results at different temperatures (Table 1), show that in poly(methyl methacrylate) and poly(methyl acrylate) the segmental motions are largely controlled by solvent flow (it is not known whether the solvent complexes with the polymer or the chromophores which were used to monitor the segmental motions).

3. Electronic Energy Migration - vinyl aromatic polymers

Polymer photophysics is dominated by excimer formation. The kinetics of the excimer formation process were described by Birks et.al.[8] according to Scheme 1. To a first approximation, any influences of diffusion or energy migration are neglected, and only the two chromophores directly involved in the excimer formation process are considered. In Scheme 1, M refers to the unexcited monomer species, $^1M*$ to the monomer in its first excited singlet state and $^1D*$ represents the excimer. k_M is the rate of depopulation of $^1M*$ by radiative or non-radiative decay to the ground state, which would be the rate of decay of $^1M*$ fluorescence in the absence of other chromophores or any intramolecular chemistry. k_M is sometimes refered to as the natural decay rate, which is normally used as a synonym to the radiative rate constant, which is the rate at which $^1M*$ fluoresces. k_M will be referred to in this report as the isolated decay rate, to make the distinction clear. k_D is the rate of radiative and non-radiative decay of the excimer. k_{DM} is the rate of formation of excimer from monomer, while k_{MD} is the rate of dissociation of the excimer to recreate the excited monomer.

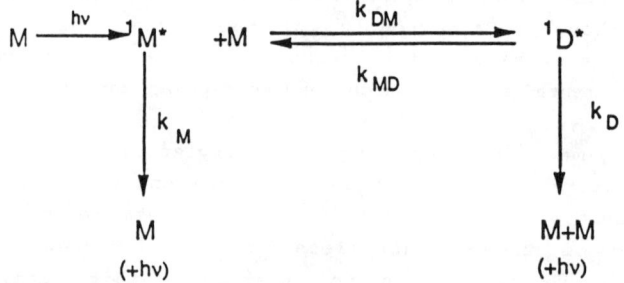

$$M \xrightarrow{h\nu} {}^1M^* + M \underset{k_{MD}}{\overset{k_{DM}}{\rightleftarrows}} {}^1D^*$$

$$\downarrow k_M \qquad\qquad\qquad\qquad \downarrow k_D$$

$$M \qquad\qquad\qquad\qquad M+M$$
$$(+h\nu) \qquad\qquad\qquad\qquad (+h\nu)$$

Scheme 1. Birks kinetic scheme

$$[^1M^*] = \frac{[^1M^*]_0}{(\lambda_2-\lambda_1)} \{(\lambda_2-X)\exp(-\lambda_1 t)+(X-\lambda_1)\exp(-\lambda_2 t)\}$$

$$[^1D^*] = \frac{k_{DM}[^1M][^1M^*]_0}{(\lambda_2-\lambda_1)} \{\exp(-\lambda_1 t)\exp(-\lambda_2 t)\}$$

where X, λ_1 and λ_2, are functions of the rate parameters k_M, k_D, k_{MD} and k_{DM};

$$X = k_M + k_{DM}[M],$$

$$\lambda_1 = 1/2\,[X+k_D+k_M-\{(k_D+k_M-X)^2+4k_{MD}k_{DM}[M]\}^{1/2}],$$

$$\lambda_2 = 1/2\,[X+k_D+k_M+\{(k_D+k_M-X)^2+4k_{MD}k_{DM}[M]\}^{1/2}]$$

Although in a few specialized cases, excimer fluorescence kinetics in synthetic polymer solutions have been shown to fit Birks kinetics[9] this is not generally the case.[10] The failure of the Birks scheme could be due to:

(a) Heterogeneous monomer sites
(b) Variations in the single energy migration length. (Figure 5)
(c) Rotational sampling of the excimer sites.
(d) Heterogeneous excimer sites.

Many groups, including our own, have expended much effort in devising mathematical models which are appropriate for description of each of the above processes, and data have been fitted to many, quite complex, mathematical functions. The work, which is beyond the scope of the brief review here, is still evolving, and there is now some cautious optimism that discrimination between mathematical models will become possible, hence permitting the relative importance of the above complications to the simple Birks scheme to be evaluated in particular circumstances.

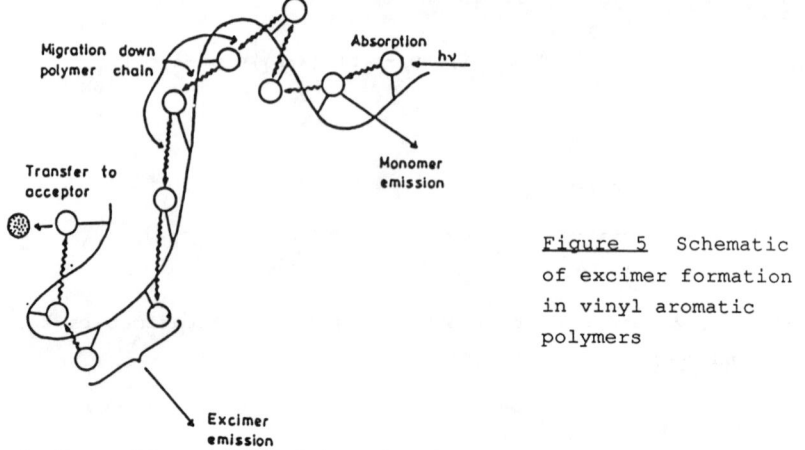

Figure 5 Schematic of excimer formation in vinyl aromatic polymers

4. Surface effects - EWIFS in poly(diacetylenes)

The solvatochromic properties of soluble polydiacetylenes have received considerable attention over the past few years, since they exhibit dramatic solvato-, thermo- and electro-chromic effects. The spectral shifts in both absorption and emission (where observable) are attributed to the degree of disorder in the conjugated π-electron backbone. In the disordered systems, such as those induced by the higher solubilities of high-temperature solutions or better solvents, the conjugation is disrupted reducing the effective conjugation length and resulting in a shift to higher energy of both the absorption and emission spectra. Conversely, a reduction in the solubility causes an increase in the effective conjugation length in the polymer backbone and results in a shift to lower energies of the absorption and emission spectra. These two cases are commonly referred to as the Y- and R-forms, respectively. In some cases the increased order in the R-form also reduces the fluorescence quantum yield to an undetectable level. The nature of the structural deformations responsible for the observed changes in spectral properties is still an area of controversy. In disordered systems it is not clear whether the deformations are regular or irregular, whilst in the ordered forms, the existence of solvated single chains is still a subject of debate. Thus the Y-form to R-form transition could be due to chain folding to form an intramolecular fringed micelle structure. This elegant idea, based on Raman data of very dilute solutions, combines both of the aforementioned ideas: whilst the transition is a single chain phenomenon, the product is an intramolecular aggregate, and at higher concentrations would lead to intermolecular aggregation with an almost identical type of structure. The conjugation lengths within the micellar structure are longer than with the disordered Y-form and hence absorb lower energy light. This interpretation also complies with conventional poor solvent effects on traditional polymers, where the polymer coils in order to minimize the interaction with the solvent.

The influence of surfaces on the properties of polymers in solution is currently receiving much interest. Total internal reflection fluorescence (TIRF) spectroscopy, or more precisely evanescent-wave induced fluorescence (EWIF) spectroscopy plays a major role in the investigation of processes that occur at a solid/solution interface. We have carried out a study[10] to determine the effect of a fused silica surface on the conformation of the soluble polydiacetylene, poly-4BCMU, in good and poor solvents. Poly-4BCMU is relatively unusual in that both the Y and R forms fluoresce. In addition to the fundamental interest in such a study, the results are also of technological importance, since it is from solution that thin films of poly-4BCMU are prepared, either by spin or dip coating, for use as planar optical waveguides.

Figure 6 shows that the surface (EWIF) spectra are identical to those of the R and Y forms measured in bulk solution, and so only two species need to be invoked to account for observed changes upon presentation of a solution of either Y form or R form to a fresh silica surface, a result shown for the Y-case in Figure 7.

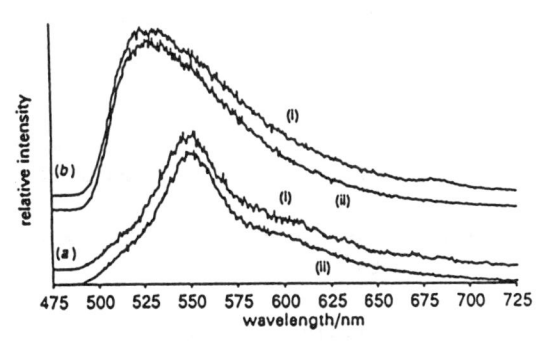

Fig. 6 Comparison of (a) R-form and (b) Y-form fluorescence spectra, measured (i) in the bulk and (ii) using EWIF

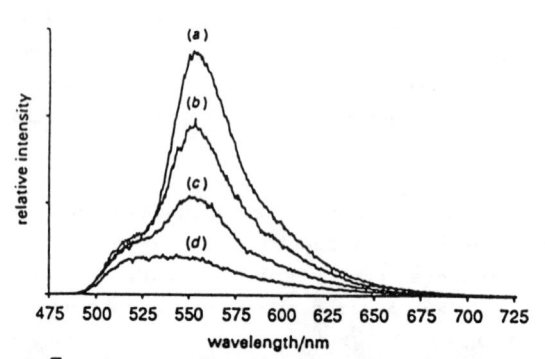

Fig. 7 EWIF spectra recorded after: (a) 2.5, (b) 20, (c) 50 and (d) 195 min, following the introduction of a Y-form solution of poly-4BCMU solution in 2-MeTHF to an air-dried fused silica surface

The results have been explained as follows.

The EWIF signal intensity is far greater than would be predicted from the bulk of the solution and has been interpreted as a decrease in the non-radiative, internal conversion deactivation mechanism for the excited state, resulting from a reduction in flexibility of the molecule adsorbed onto the glass surface. This mechanism does not lead to fluorescent poly-4BCMU crystals, because of the existence of efficient energy transfer to non-fluorescent traps in the crystal.

The time-dependent spectral changes are a result of a transient response of the polymer solution encountering a 'non-solvated' or hydrophilic glass surface. In the case of a Y-form polymer solution the surface causes the dynamic equilibrium between the Y-form and R-form to favour the ordered R-form, which then adheres to the surface. The system returns to equilibrium by the conditioning or 'solvating' of the surface, causing the R-form to revert back to the Y-form, which continues to adhere to the surface. This result may be of importance to the waveguiding properties of spin- or dip-coated poly-4BCMU waveguides, prepared from untreated glass substrates.

The rate at which the equilibrium between the R-form and Y-form is disturbed supports the intramolecular fringed micelle structure for the R-form. The possibility of the formation of aggregates is not ruled out, since the proposed structure is consistent with the aggregates forming a similar intermolecular fringed micelle structure. It is therefore proposed that the R-form can exist both in a single chain and, if the concentration is sufficiently high, in aggregates also.

Acknowledgements

This paper has drawn upon the work of A.J. Roberts, G. Rumbles, R.C. Drake, C.F.C. Porter, R.L. Christensen, A.J. Brown and M.J. Carey, of IC, and Professor Ian Soutar and his group, University of Lancaster. All are thanked for their contribution. Financial support from SERC and The Royal Society is gratefully acknowledged.

References

1. A.C. Somersall and J.E. Guillet, *J. Macromol. Sci. Rev. Macromol. Chem.*, 1975, C13, 135.
2. D. Phillips, *British Polymer J.*, 1987, 19, 135 (and references therein).
3. D.V. O'Connor and D. Phillips,'Time-Correlated Single Photon Counting', Academic Press, London, 1984.
4. M. Graley, A. Reiser, A.J. Roberts and D. Phillips, *Macromolecules*, 1981, 14, 752.
5. G. Rumbles and D. Phillips in 'Applications of Lasers in Polymer Science and Technology', Eds. J.P. Fouassier and J. Rabek, CRC Press, 1990, 1, 91 (and references therein).
6. R.L. Christensen, R.C. Drake and D. Phillips, *J. Phys. Chem.*, 1986, 90, 5960.
7. R.C. Drake, Ph.D. Thesis, University of London, 1986.
8. J.B. Birks, D.J. Dyson, and T.A. King, *Proc. Roy.Soc.*, 1964, A277, 270.
9. C.F.C. Porter, Ph.D. Thesis, University of London, 1990.
10. G. Rumbles, A.J. Brown, D. Phillips and D. Bloor, *JCS Faraday Trans.*, 1992, 88, 3313 (and references therein).

Third-order Nonlinear Optical Processes in Conjugated Polymers – From Materials to Waveguide Devices

W. J. Blau

DEPARTMENT OF PURE AND APPLIED PHYSICS, TRINITY COLLEGE, UNIVERSITY OF DUBLIN, DUBLIN 2, IRELAND

Abstract. Conjugated polymers possess a large off-resonant nonlinear optical response with an ultrafast relaxation time. In the near-infrared they show enhanced response due to multi-photon resonances. By comparing optically similar, but chemically different materials, qualitative structure-property relationships can be derived which lead to improved material response. Using conventional thin film deposition techniques known from microelectronics, optical waveguiding structure can be fabricated, which may lead to the development of polymeric all-optical switching devices.

1. INTRODUCTION

Since the first planned synthesis of a conjugated polymer in the form of single crystalline polydiacetylene by Wegner in 1969 [1], a range of semiconducting polymers has been developed with very intense absorption bands in the visible or near-infrared wavelength region. This originates from the delocalised π-electron system in the polymer backbone. Already in 1976 a French-American team [2] reported on the observation of optical nonlinearities in polydiacetylene measured by frequency tripling of near-infrared laser pulses. In this report the authors point out that the observed nonlinearities are larger than the values observed non- resonantly in conventional inorganic semiconductors. Obviously a broadband nonlinear response in this wavelength region is interesting for optical communication systems that operate around 1.3 and 1.55 µm. To date, nonlinear response has been observed in almost all conjugated polymers. Technical applications, however, rely on processable materials [3] which can be developed into thin film devices [4].A large number of inter-disciplinary, both university and industry-based research groups are active in this area now, mainly in Europe, USA and Japan [5,6]. In this introductory article the author intends to introduce the basic ideas underlying the origins and use of nonlinear optical effects in conjugated polymers, using examples obtained in his laboratory.

Figure 1 shows the chemical structural formulae and optical absorption spectra of 3 exemplary polymers whose response will be discussed below: polydiacetylene, polythiophene and poly (paraphenylene-ethinylene). By attaching long aliphatic side chains these polymers are made soluble in organic solvents and hence can be solution processed. Measurement of their properties in solution will give information about the molecular response of single (quasi) isolated chains. By increasing the concentration intermolecular affects can be quantified until eventually the solid-state properties are reached.

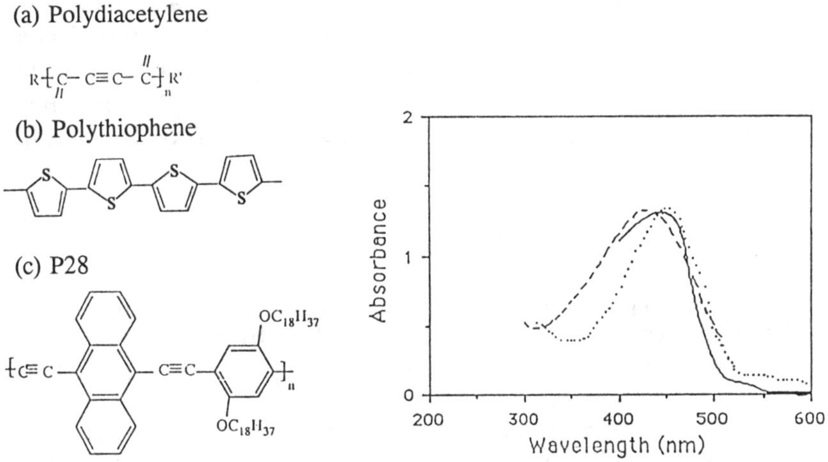

(a) Polydiacetylene

(b) Polythiophene

(c) P28

Figure 1. Chemical structural formulae and absorption spectra of the selected conjugated polymers: (a) (4BCMU) - Polydiacetylene (—), (b) (3-Butyl) - Polythiophene (---), and (c) Poly (paraphenylene ethinylene) (···).

2. THIRD-ORDER OPTICAL NONLINEARITIES: DEFINITION AND MEASUREMENT

In classical optics the interaction of light with matter is characterised by material parameters such as absorbance, refractive index and scattering. All of these are wavelength dependent, but independent of the intensity I of the light source used in the measurement. High intensity light, produced for instance by a short pulse laser, can modify these properties and will cause a nonlinear optical response [7]. An incident light field E induces a dipole moment P_{ind} in a material which can be (approximately) described by a power series

$$P_{ind} = \varepsilon_0 \chi E$$

$$= \varepsilon_0 (\chi^{(1)} + \chi^{(2)}E + \chi^{(3)}E^2 + ...)\cdot E \qquad (1)$$

The overall susceptibility χ is linked to the complex refractive index n_c by

$$n_c^2 = 1 + \chi \qquad (2)$$

The constant $\chi^{(1)}$ is the "classical" linear susceptibility, $\chi^{(2)}$ and $\chi^{(3)}$ are the second- and third-order nonlinear susceptibilities observed, e.g. by frequency doubling and tripling.

Due to material symmetry restrictions, the second-order susceptibility vanished in centrosymmetric materials and hence $\chi^{(2)} = 0$ holds in (most) conjugated polymers. The lowest observable nonlinearity is therefore $\chi^{(3)}$. As the intensity I is proportional to E^2, $\chi^{(3)}$ can also describe an intensity dependent complex refractive index. In SI units the nonlinear refractive index $n_2 = \Delta n/I$ relates to $\chi^{(3)}$ by

$$n_2 = 3\, \mathcal{R}e\, \chi^{(3)} / (4\, \varepsilon_0\, c\, n_0^2) \qquad (3)$$

The imaginary part of $\chi^{(3)}$ relates to light induced absorption changes. These can either occur as saturable absorption in resonant cases ($Im\, \chi^{(3)} < 0$) or as two- or multi-photon absorption ($Im\, \chi^{(3)} > 0$) in transparent materials.

When comparing molecular systems it is often useful to calculate molecular nonlinearities - so called hyperpolarisabilities; β, γ, ... - from the macroscopic $\chi^{(3)}$ by normalising to the number density of molecular (monomeric) species N and taking into account local electric field factors F:

$$\gamma = \chi^{(3)} \cdot (N \cdot F)^{-1} \qquad (4)$$

A number of experimental techniques are available for the measurement of $\chi^{(3)}$; most commonly applied is third-harmonic generation. The technique used to obtain the results described below however relies on a direct observation of the refractive index changes via a laser-induced grating (LIG) [8]. Two coherent laser pulses are overlapped at a small angle ($\sim 2°$). The interference fringe pattern generated will, through the nonlinearity, induce a local refractive index pattern resembling this fringe pattern. This is commonly termed a laser induced grating or real-time hologram. By diffracting one of the writing beams or a third laser pulse off this

LIG the magnitude and time response of $\chi^{(3)}$ is determined [9]. If more than one photon is required per writing pulse, i.e. if two or three photon processes dominate, this will be directly observable through the intensity dependence of the process. For nonresonant or one photon resonant cases the diffraction efficiency of the LIG will be proportional to I^2, and $\chi^{(3)}$ becomes a material constant. For two and three photon resonances however, $\chi^{(3)}$ becomes intensity dependent and the diffraction efficiency will depend on the intensity as I^4 or I^6 respectively.

Finally it should be stressed that in polymers $\chi^{(3)}$ and γ can possess both a real and an imaginary part. By comparing the polymer response in solution at different concentrations to the real and positive nonlinearity of a typical organic solvent, the magnitude of both parts and the sign of the real part can be determined. At high concentrations aggregation effects will also appear.

3. COMPARISON OF EXPERIMENTAL RESULTS OF SELECTED POLYMERS IN SOLUTION

In this section a series of exemplary LIG measurements is described using infrared picosecond laser pulses at $\lambda = 1.064$ µm. As seen in Figure 1 the polymers selected for this discussion show approximately the same absorption spectra. The electronic structure originates generally from an interband transition in the one-

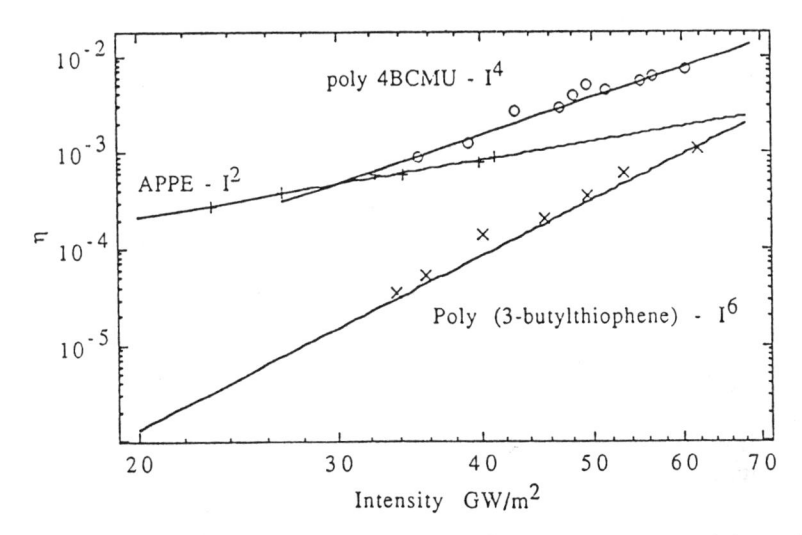

Figure 2. Intensity dependence of the diffraction efficiency of the LIG at 1.064 µm for the 3 samples shown in Figure 1.

dimensional semiconductors [10]. The intensity dependence of the diffraction efficiency of the LIG however shows marked differences, as seen in Figure 2.

Only the poly(paraphenylene-ethinylene) shows a square dependence, characteristic of $\chi^{(3)}$ = const. In poly (4-butoxycarbonyl methylurethane) diacetylene (p-4BCMU) a two photon resonance dominates. Such a resonance can also be observed by third-harmonic generation with a tunable laser [11]. Polythiophene shows even an I^6 dependence, typical of a three-photon resonance. It should be noted that in all polymers a significant imaginary part of the nonlinearity (up to 60%) is observed, indicative of dissipative multi-photon absorption. This observation highlights the importance of electron-vibration coupling in such low-dimensional π-electron systems [12] and also electron-electron correlation [13]. Recently Grossman et al. [14] have presented calculations on polyenes which take such correlations into account. They predict a scaling of the nonlinearity with polymer chain length as $\chi^{(3)} \approx L^{4.6}$. In fact, experimental measurements using various oligomers of different well-defined lengths confirm such a scaling law within experimental errors [15,16] up to a length of L ≈ 70 Å. This length limitation, in turn, agrees with electron-vibration coupling as the limiting factor for the extent of π-electron delocalisation. Therefore a strong imaginary component of the nonlinearity will appear. When introducing transition metal atoms into the polymer backbone, these interactions can be systematically modified and the nonlinearity of nickel containing diacetylenes and thiophenes is enhanced by more than an order of magnitude [17] over their purely organic analogues.

4. OPTICAL WAVEGUIDING IN POLYMER FILMS

Generally the light intensities necessary to observe strong nonlinear optical effects ($\approx$ GW/cm^2) cannot be obtained from diode lasers such as those used in optical communication systems. In optical monomode waveguide, however, the interaction area (~10 μm^2) is physically limited. Thus, by choosing a device length of up to several mm, nonlinear devices which operate with diode lasers might be constructed. Polydiacetylene, for example, (most notably p-4BCMU) can be made into a planar thin film waveguide by spin coating from solution onto glass substrates in clean room conditions. Definition of a waveguiding channel is then obtained by removing the unwanted polymer by excimer laser photoablation [18] or by photochemical reactions [19]. In the research laboratory coupling of light into and out of the waveguides is obtained from prism coupling using SF6 glass prisms mounted on a precision rotation table. Waveguiding is observed as dark m-lines or by visual observation of the light scattered in the guide. In poly (3BCMU) diacetylene waveguiding was observed with visible and infrared laser diodes and a lower power Nd-glass laser. In a planar 2.7 μm thick film as example, 3 TM modes are observed at 670 nm and 1 TM and 5 TE modes are seen at 860 nm [20]. Waveguide losses are measured from the length dependence of the light scattered in the waveguide. For this measurement the scattered light is collected with a 200 μm core glass fibre scanned along the guide and detected by a Si photodiode. The decay of the scattered light intensity along the guide is recorded. Attenuation is generally measured in dB/cm and obtained from the slope of the measured curve as shown in Figure 3. Between 670 nm and 1300 nm the attenuation coefficient decreases from 20dB/cm to 4dB/cm. Such attenuation values appear feasible for first device prototype demonstration.

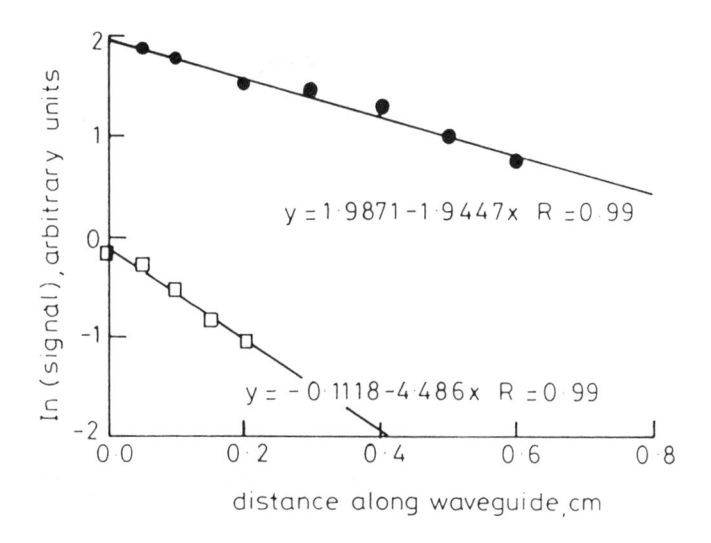

Figure 3. Measurement of waveguide attenuation in (3BCMU) Polydiacetylene by scanning a pigtailed detector along the guide. The attenuation of the guide is obtained from the slope of the graph and varies with wavelength (Squares: λ = 670 nm; Attn = 20 dB/cm; Circles: λ = 860 nm; Attn = 9 dB/cm).

Nonlinear optical waveguide response can be observed in various ways, for example by observing the intensity dependence of the optimum prism coupling angle. The coupling angle shifts due to the intensity dependent film refractive index and a nonlinear susceptibility of $\chi^{(3)} = 4 \cdot 10^{-19}$ m^2/V^2 is observed at λ = 1064 nm for (3BCMU) Polydiacetylene. The dispersion of the nonlinearity is shown in Figure 4 in the near infrared. The position of the two-photon resonance can be clearly seen as enhancement of $\chi^{(3)}$ at approximately twice the main electronic absorption peak wavelength (620 nm). The exact position of the two-photon resonance has been a subject of much discussion over the past decade as it provides spectroscopic information about the position of one-photon forbidden energy levels.

A simple example for an all-optical switch is the LIG technique described above. Only if both writing beams and reading beams are present and have the proper wavelength a diffracted signal is observed. Hence such an arrangement constitutes a time and

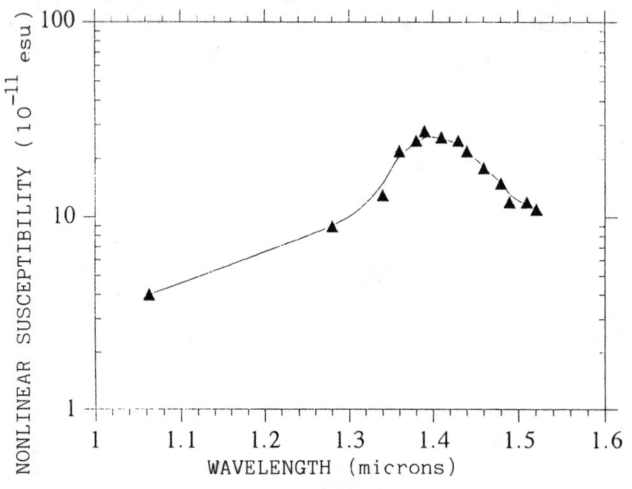

Figure 4. Dispersion of the optical nonlinearity in (3BCMU) Polydiacetylene in the infrared. Please note that the unit for $\chi^{(3)}$ used traditionally in this field is the "esu" which relates to the metric system by $\chi^{(3)}$ (SI) / $\chi^{(3)}$ (esu) = $4\pi \cdot 10^8/c^2$.

wavelength- dependent optical AND gate. Indeed, such a device can be fabricated in a waveguiding geometry and a diffraction efficiency of 15% is seen with 2 W laser peak power. As the relaxation time of the nonlinearity is <2 ps, the bandwidth of such a switch exceeds 300 GHz. Its speed power products is 5pJ which is comparable to a conventional electronic TTL gate. An open question however remains the chemical and thermal stability of such a polymer device in permanent use in a communication system.

Acknowledgements
The author would like to thank all present and past members of his research group for their work and enthusiasm, in particular Dr. H.J. Byrne and B. Rossi (in memoriam). Part of this research is supported by the Commission of the European Communities through the project RACE 1020/2012.

References

1. G. Wegner: Zeitschr.Naturforsch. **24B**, 824 (1969).
2. C. Sauteret, J.P. Hermann, R.Frey, F. Pradere, J. Ducuing, R.H. Baughman, R.R. Chance: Phys. Rev. Lett. **36**, 956 (1976).
3. G.N. Patel: Polym. Prepr., Am. Chem. Soc., Div. Polym. Chem **19**, 154 (1978).
4. P.D. Townsend, G.L. Baker, N.E. Schlotter, C.F. Klausner, E. Etemad: Appl. Phys. Lett. **53**, 1782 (1988).
5. D.S. Chemla, J. Zyss (Ed.): "Nonlinear Optical Properties of Organic Materials and Crystals", vol. **2**, (Academic Press, Orlando, (1987)).
6. J. Messier, F. Kajzar, P. Prasad, D. Ulrich (Ed.): "Nonlinear Optical Effects in Organic Polymers" (NATO ASI Series 162, Kluwer, Dordrecht, (1989)).
7. Y.R. Shen: "The Principles of Nonlinear Optics" (Wiley, New York, (1984)).
8. H.E. Eichler, P. Günter, D.W. Pohl: "Laser Induced Dynamic Gratings" (Springer, Berlin, (1986)).
9. P. Horan, W. Blau, H. Byrne, P. Berglund: Appl. Opt. **29**, 31 (1990).
10. D. Bloor, R.R. Chance (Ed.): "Polydiacetylenes" (NATO ASI Series 102, Nijhoff, Dordrecht, (1985)).
11. F. Kajzar, J. Messier, in Ref. 10, p. 325.
12. H.J. Byrne, W. Blau: Synth. Met. **37**, 231 (1990).
13. B.S. Knudsen, B.E. Kohler: J. Chem. Phys. **64**, 4422 (1976); Z.G. Soos, S. Ramasesha: Phys. Rev. **B29**, 5410 (1984).
14. C. Grossman, J.R. Haflin, K.Y. Wong, O. Zamani-Khamiri, A.F. Garito, in: Ref. 6, p. 61.
15. H.J. Byrne, W. Blau, R. Giesa, R.C. Schulz: Chem. Phys. Lett. **167**, 484 (1990).
16. P.N. Prasad, in: Ref. 6, p. 351.
17. W. Blau, H.J. Byrne, D.J. Cardin, A. Davey: Journ. Mater. Chem. **1**, 67 (1991).
18. J.D. Magan, P. Lemoine, H.J. Byrne, W. Blau: Journ. Molec. Electr. **5**, 247 (1989).
19. K. Rochford, R. Zanoni, Q. Gong, G.E. Stegeman: Appl. Phys. Lett. **55**, 1161 (1989).
20. B.Rossi, W. Blau, J. O'Gorman, S. König, D. Westland, V. Skarda: Electron. Lett. **26**, 312 (1990).

Polymer/CT Salt Composites as $\chi^{(3)}$ Media

G. H. Cross,[1*] M. Carroll,[1] T. L. Axon,[1] D. Bloor,[1]
R. Rangel-Rojo,[2] A. K. Kar,[2] and B. S. Wherrett[2]

[1] DEPARTMENT OF PHYSICS, APPLIED PHYSICS GROUP, UNIVERSITY OF DURHAM, DURHAM DH1 3LE, UK
[2] DEPARTMENT OF PHYSICS, HERIOT-WATT UNIVERSITY, EDINBURGH EH14 4AS, UK

ABSTRACT

Polymeric composites containing microcrystals of 1:1 salts of tetracyano-p-quinodimethane, TCNQ, have been prepared. Recrystallisation of these organic one-dimensional semiconductors in-situ by either heat treatment or exposure to vapour yields uniform dispersions of crystallites of sub-optical dimensions. We report a strong dependence of the lowest energy electronic absorption band in the lithium, sodium and potassium TCNQ salt with crystallite size. The inter-molecular charge transfer band shows a shift of up to 0.4 eV between films where the crystallite size is a few microns and those where crystals cannot be observed optically. The non-linear refraction of a heat-treated film containing TEA$^\oplus$(TCNQ)$^\ominus$ (TEA$^\oplus$ - triethylammonium) is reported. $\chi^{(3)}$ at 712 nm is 2.69 x 10^{-10} esu and is found to be a bleaching resonant non-linearity.

1. INTRODUCTION

There has been little gain in real terms in the third order non-linearity, $\chi^{(3)}$, of organic materials since the resonant enhanced value of 8.5 x 10^{-10} esu for the polydiacetylene, PTS, was reported in 1976 [1]. This is despite widespread efforts in novel synthesis and a growing understanding of the basic molecular requirements. Larger values in new materials are invariably associated with a concomitant increase in absorption, whether it be a one or two photon process. If the problems of gaining a clear advantage in the trade off between non-linearity and absorption are to be overcome, then what may be termed novel 'engineering' approaches need to be coupled with new materials synthesis.

One such example is in the fabrication of quantum confinement semiconductor structures [2,3]. In the non-resonant case these act to increase the non-linearity either through confinement induced anharmonicity in the exciton energy [2] or through local field enhancement in dielectric microcavities [4,5]. In the resonant case, the enhancement results from a transient reduction in the exciton oscillator strength following the crystallite surface trapping of a photo-induced electron-hole pair [6].

A further solution to the problem of finding a viable trade off between $\chi^{(3)}$ and absorption coefficient, α, is to operate in a near-resonance frequency region and tailor the degree of detuning from the absorption edge. In the long wavelength tail of a dominant optical transition one can expect to find a frequency where the ratio, $\chi^{(3)}/\alpha$, is optimised. This principle lies behind some recent work on metal dithiolenes and dithiolates [7,8] whose absorption band is tunable in the region 1500 nm to 700 nm. Step-wise control of the absorption characteristics is obtained by varying the substituent groups around the metal-centred chromophore. Finding an optimum near resonance operating point when the laser wavelength must be held fixed then requires multiple chemical modifications.

A rather more effective method might be to tune the absorption characteristics in a continuous manner using _physical_ modifications. A system demonstrating the feasibility of this approach, lies with some organic charge transfer, CT, salt/polymer composites we have fabricated [9]. The initial studies have centred around the 1:1 salts of TCNQ where the counter cation is an alkali metal or the cation triethylammonium, TEA. We present here our work on these materials and report the fabrication of microcrystal dispersed polymer thin films whose optical properties are dependent on the crystal size. Some preliminary studies on the non-linear optical properties of films containing $TEA^{\oplus}(TCNQ)^{\ominus}$ are also presented.

2. EXPERIMENTAL

TCNQ salts of $Li^{\oplus}$, $Na^{\oplus}$, $K^{\oplus}$ and triethylammonium, $TEA^{\oplus}$ were prepared according to the method of Melby et al. [10] and the stoichiometry was verified by elemental analysis and by the method of UV/vis. spectroscopy as solutions in acetonitrile [11]. The latter method gives the stoichiometric ratio of $TCNQ^{\ominus}$ to neutral TCNQ in prepared salts and is a useful check on the TCNQ composition of salts following heat treatment in the polymer film. The salts were co-dissolved with PMMA in acetonitrile at a composition of between 1.5 and 3.5 weight percent and thin films were then prepared by substrate withdrawal from solution.

The molecularly dispersed polymer films formed via this process were then treated to induce crystallite formation. (i) Vapour Treatment. Following the general approach of Kita et al. [12] for reticulate doping of polymers with CT salts, we expose the film to solvent vapour and observe the conversion to microcrystallites visually as distinct colour changes. (ii) Heat Treatment. The films may be converted by heating to above the glass transition temperature of the polymer.

During vapour treatment, the 'as prepared' film, which we assume is a solid ionic solution as evidenced from linear absorption studies, is held over solvent in a closed vessel for periods of up to a few hours. Where crystallisation occurs, the green films slowly change colour to give a blue coloration and UV/vis. spectroscopy can confirm the efficiency of conversion (*vide infra*). Heat treatment of films is simply a matter of heating in a vacuum oven for 1 hour at 150°C where, without exception in the series of materials we have examined, the films convert to a deep blue coloration and are optically clear (non-scattering) to the eye.

Degenerate four wave mixing, DFWM, studies at 712 nm, i.e. on resonance, were performed using pulses of 0.7 ps duration. $\chi^{(3)}$ is determined by measuring the intensity of the phase conjugate signal in reflection from the sample and the decay times are measured by delaying the back pump beam. To complement these studies, we use a pump-probe measurement, to make an independent measurement of $\text{Im}\chi^{(3)}$. We also have undertaken both open and closed aperture Z-scan measurements using Gaussian profile pulses of a few picoseconds duration at 532 nm. This wavelength is to the high energy side of the lowest energy electronic excitations in the films.

3. RESULTS AND DISCUSSION

3.1 Vapour Treated Films

Vapour treatment gives results strongly dependent on the vapour used. The principle of crystal formation relies on the swelling of the host medium on exposure to vapour. This allows increased mobility of the dissolved ionic components which can then associate and crystallise in-situ. The dynamics of the recrystallisation process depend on the partitioning, between immobile and mobile phases, of the ionic moieties (TCNQ$^\ominus$ and alkali metal counter ions). It follows that those materials which are less soluble in the solvent vapour favour rapid crystallisation by this method. The variability in crystallite formation is most pronounced in the case of Li$^\oplus$(TCNQ)$^\ominus$. This salt is one of the most soluble of the TCNQ salts in solvents such as acetone or acetonitrile whereas it is insoluble in dichloromethane or chloroform. The UV/vis.

absorption spectra of films containing Li⊕(TCNQ)⊖ following vapour treatment are shown in Figures 1(a) and (b).

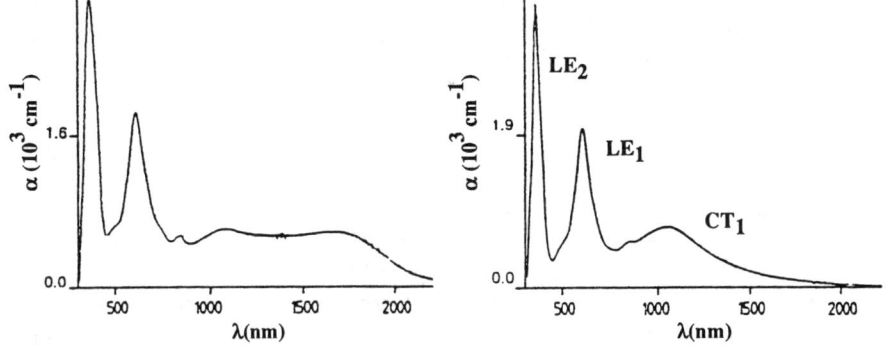

Figure 1(a): Linear absorption of Li⊕(TCNQ)⊖ in PMMA following treatment with acetonitrile vapour.

Figure 1(b): Linear absorption of Li⊕(TCNQ)⊖ in PMMA following treatment with chloroform vapour.

Vapour treated films of lithium TCNQ are clear blue in colour directly as a result of the strong excitations LE_1 and LE_2 which are readily identified in Figure 1(b). These bands have been assigned to local or, intramolecular, excitations in TCNQ [13,14]. The CT_1 band in these films arises as a result of charge excitation along the TCNQ stack and lies at 1.2 eV.

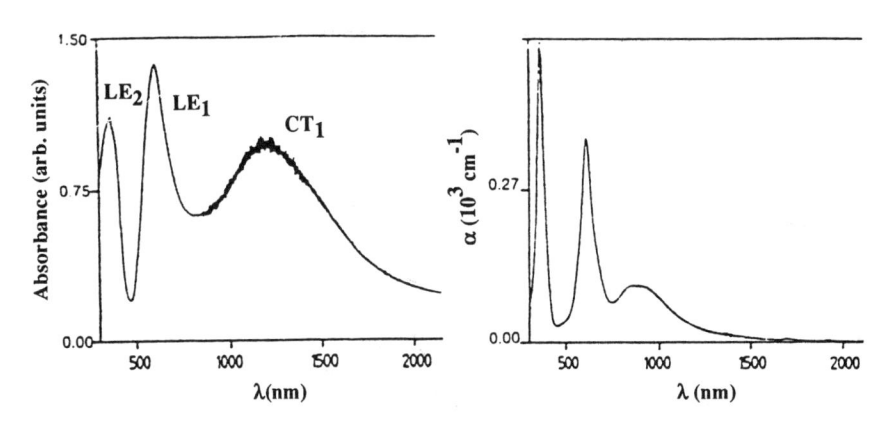

Figure 2(a): Linear absorption spectrum of K⊕(TCNQ)⊖ in KBr disc.

Figure 2(b): Linear absorption spectrum of K⊕(TCNQ)⊖ in PMMA following treatment with dichloromethane vapour.

The spectra of the alkali metal TCNQ salts dispersed in KBr represent the excitations present in bulk crystals. Thus the spectrum of a KBr disc containing K$^\oplus$(TCNQ)$^\ominus$ is shown in Figure 2(a) to identify all the major charge transfer intermolecular and intramolecular excitations in TCNQ salts where the TCNQs form a regular stack and where the band filling is one half.

This spectrum can now be compared to that for a K$^\oplus$(TCNQ)$^\ominus$ loaded film which has been vapour treated using dichloromethane, a non-solvent for the salt (Figure 2(b)). The bands LE$_1$ and LE$_2$ are unchanged between the samples but notably, the CT$_1$ band is considerably blue shifted, lying at around 1.37 eV. Examination of films of K$^\oplus$(TCNQ)$^\ominus$ under an optical microscope reveal no discernible crystallite phase; thus crystals are of a much smaller dimension than optical wavelengths or fortuitously have the same refractive index as the host (n ~ 1.49), which is unlikely.

These features are repeated in the case of sodium TCNQ, whose KBr spectrum and spectrum as a microcrystalline dispersed salt in PMMA is shown in Figure 3.

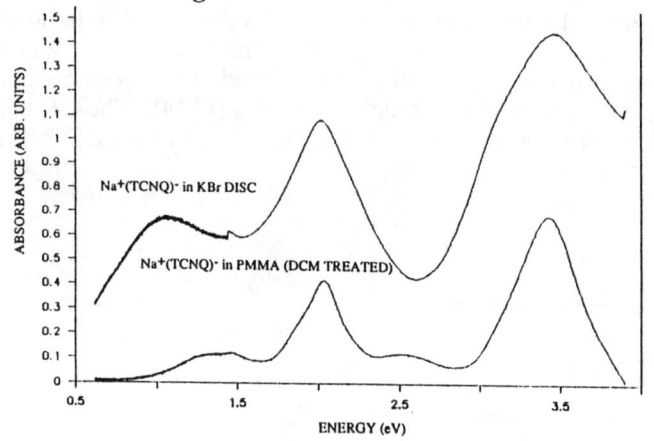

Figure 3: Na$^\oplus$(TCNQ)$^\ominus$ in KBr disc (upper trace) and dispersed into PMMA film following treatment with the vapour of dichloromethane.

3.2   Heat Treatment

We have found that heating the films to a point above the glass transition temperature of the polymer (105°C) produces a rapid re-crystallisation of the salt. The resulting films contain sub-micron scale microcrystals with spectra reminiscent of those for vapour treated films. The shifts in the CT$_1$ band are larger however, and are around 0.4 eV.

Table 1 summarises the positions of this band with the salts contained in the various host media and following either solvent vapour or heat treatment. The data for Rb⊕TCNQ⊖ and the values for the TCNQ dimer intermolecular transfer integral, t, are taken from reference [13]. The overall position of the CT$_1$ band is proportional to the energy of overlap between adjacent TCNQ's in the linear chain and in the case of these 1:1 salts, on the degree of electron-electron correlation.

SALT	λ_{max}, KBr	λ_{max}, vapour	λ_{max}, heat	t(eV)
Li⊕TCNQ⊖	1126 (1.09 eV)	1025 (1.2 eV)		-
Na⊕TCNQ⊖	1170 (1.05 eV)	900 (1.37 eV)	830 (1.48 eV)	0.268
K⊕TCNQ⊖	1204 (1.02 eV)	900 (1.37 eV)	862 (1.43 eV)	0.271
Rb⊕TCNQ⊖	1176 (1.05 eV)	-	-	0.313

Table 1: Details of the position of the CT$_1$ band absorption maximum for crystals of the 1:1 TCNQ salts in KBr and in polymer films following either solvent vapour (non-solvents) or heat (at 150°C) treatments. t is the dimer overlap integral taken from reference 13.

Table 1 shows nearly identical behaviour between the K⁺ and Na⁺ salts of TCNQ which is in line with their similarity of structure. Each material forms a chain of TCNQ's comprising a weak dimerisation and with a staggered ring-ring overlap between TCNQ's [15,16]. We are at present preparing samples of Rb⊕TCNQ⊖ whose structure is one of a linear chain where TCNQ's stack with ring-external double bond overlap [17].

The lithium salt seems to fall outside the scheme shown by the alkali metal series. Although the KBr disc spectrum shows CT$_1$ at 1.09 eV and thus in a similar position to the others, solvent treatment with acetonitrile, as previously discussed produces a red shift in the band.

Figure 1(a) is of an acetonitrile vapour treated film following exposure over a 15 minute period. If the film is left exposed for 16 hours however, the spectrum shown in figure 4 is returned. The CT$_1$ band maximum is at 1600 nm (0.77 eV). Large microcrystals of dimension 10 microns or greater are formed in this film.

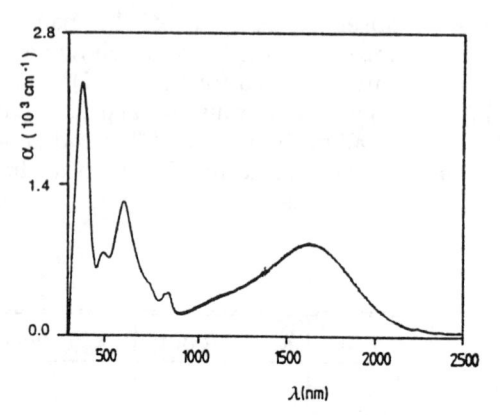

Figure 4: Acetonitrile vapour treated film containing Li⊕TCNQ•. The exposure time was 16 hours and macroscopic scale (> 10 microns) crystals are produced.

Li⊕TCNQ• is the only one of our series for which there is no crystal structure information. This is presumably due to the difficulty of preparing large scale crystals. We are currently working to determine the crystallite sizes of this and the other salts when dispersed into KBr and into PMMA. One suggestion for the apparent anomaly in the direction of the band shift is that the particles of Li⊕(TCNQ)• in the KBr disc are already of a small enough size to produce the blue-shifts in the spectra and that the film spectra, where the particles are certainly greater than 10 microns, are representative of 'bulk' Li⊕TCNQ•.

3.2.1 Heat Treated Films of TEA⊕(TCNQ)$_2$• and TEA⊕(TCNQ)•

Heat treatments to PMMA solid solutions of either the 1:2 or 1:1 salts of triethylammonium, TEA⊕, give identical spectra and the point of most significance is that in neither film is it possible to identify the CT$_1$ or CT$_2$ transitions (see Figure 5). Neither do we observe microcrystallites and we propose instead that the anions and cations form associated pairs in PMMA solution without necessarily forming crystals. There are two observations which preclude the possibility of serious degradation of the salt components. Firstly, we can re-dissolve heat treated films of the TEA⊕ salts in acetonitrile and observe the solution spectra. These show the expected TCNQ• bands at 840 nm and 395 nm and the peak height ratio [11] (A_{840}/A_{395} - 0.6, A is the absorbance value) confirms that only TCNQ• is present in the films. In the case of TEA(TCNQ)$_2$ films we assume that either the neutral TCNQ has been ionised via impurities or has been removed from the film by sublimation. Secondly, the sublimed product formed on the substrate above the film following heat treatment has an identical absorption spectrum to the 1:1 salt dispersed in a KBr disc and is thus unambiguously assigned as TEA⊕(TCNQ)•.

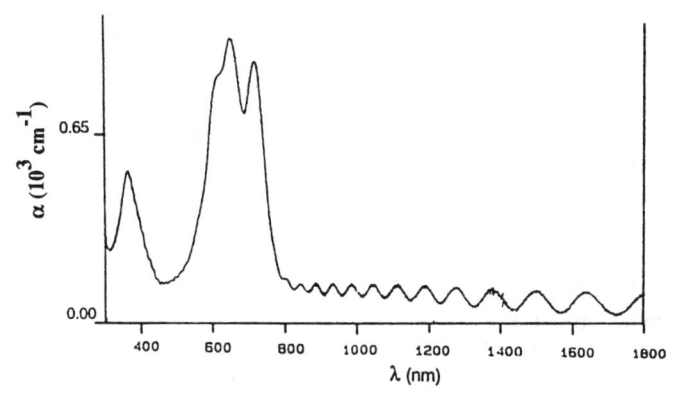

Figure 5: Linear absorption spectrum of TEA$^⊕$(TCNQ)$_2$$^⊖$ (notionally) in PMMA following heat treatment

To account for the strong absorption bands at 1.72 eV (720 nm) and 1.90 eV (650 nm) we propose a Donor-Acceptor charge transfer process. If this is so, then there are two possible configurations for the salt in heat-treated PMMA films:

(i) Ion pairs in solid solution.

(ii) Crystals of 'mixed stack' material.

3.2.2 Nonlinear Optical Studies

Some preliminary results from a study by degenerate four wave mixing on TEA$^⊕$(TCNQ)$^⊖$ heat-treated films shows a moderate nonlinearity either on (at 712 nm) or near (at 532 nm) resonance. Table 2 summarises our findings. At 712 nm using sub-picosecond pulses we have identified two excitation lifetimes of τ_1 = 6 ps and τ_2 = 75 ps. The origins of this behaviour need further examination.

The pump probe experiments at 712 nm show a similar temporal decay behaviour and from the peak values of $\Delta T/T$ we calculate a bleaching resonant non-linearity corresponding to $\text{Im}\chi^{(3)}$ = -2.75 x 10^{-10} esu. Thus in this strong one-photon resonant region, the non-linearity is nearly entirely imaginary and involves bleaching of the transition.

The z-scan measurements cannot give time resolved information but by measuring the transmitted beam intensity in closed aperture and open aperture formats, $\text{Re}\chi^{(3)}$ and $\text{Im}\chi^{(3)}$ are simultaneously obtained. For these TEA$^⊕$(TCNQ)$^⊖$ films we obtain, at 532 nm;

$\text{Re}\chi^{(3)}$ = 1.098 x 10^{-10} esu, $\text{Im}\chi^{(3)}$ = -1.715 x 10^{-11} esu, and from the linear absorption data, α = 10^2 cm^{-1}.

λ(nm)	α(cm^{-1})	$\chi^{(3)}$(esu)	Re$\chi^{(3)}$/I$_m\chi^{(3)}$	$\chi^{(3)}$/α
712	1.2 x 10^3	2.69 X 10^{-10}	~ 0	2.24 x 10^{-13}
532	10^2	1.111 X 10^{-10}	6.40	1.11 x 10^{-12}

Table 2: The linear absorption and nonlinear optical properties of PMMA films containing 1 % w/w of TEA$^⊕$(TCNQ)$^●$

We might compare these results with those of some semiconductor-doped glasses recently reported [18]. An optical filter glass containing CdS$_{0.12}$Se$_{0.88}$ measured by DFWM at 580 nm has $\chi^{(3)}$/α = 4 x 10^{-11} esu cm. The II-VI semiconductor microcrystallites also exhibit bi-exponential decay characteristics of very similar decay time as those found here.

4. CONCLUSIONS

We have developed and presented techniques whereby microcrystals of sub-optical dimensions may be uniformly dispersed in a polymeric thin film. Large blue shifts have been observed in the intermolecular TCNQ$^●$ → TCNQ$^●$ transition in K$^⊕$(TCNQ)$^●$ and Na$^⊕$(TCNQ)$^●$ microcrystals in the film compared to the 'bulk' material. This is tentatively associated with changes to the bulk electronic properties when the crystallites are of sub-optical dimensions. Li$^⊕$(TCNQ)$^●$ shows a red shift from its KBr spectrum when crystals of 10 microns or more are produced in PMMA films. We believe that the bulk spectral properties for this material are not returned by simply dispersing the powder into the KBr disc. Crystalline habit in vapour treated films is seen to be influenced by the partitioning of the solute material between immobile and mobile phases. Films of TCNQ salts of the alkali metals K$^⊕$, Na$^⊕$ and Li$^⊕$ form uniformly dispersed microcrystals of the corresponding 1:1 simple salts. Films containing TEA$^⊕$(TCNQ)$^●$ do not form segregated stack microcrystals. These form mixed stack crystals or solvated ion pairs dispersed in PMMA and give non-linearities, near resonance, around one order of magnitude lower than some of the semiconductor-doped glasses which are under scrutiny at present. We have also measured a purely imaginary bleaching non-linearity on resonance at 712 nm. further measurements of interest in this system will be to examine how the ratio Re$\chi^{(3)}$/Im$\chi^{(3)}$ varies in the wavelength region to the low energy side of the strong one-photon transitions. The convenience of the thin film format for studies will widen the possibilities for waveguide measurements in the off-resonant regions. The TCNQ salts, as a family, will provide an excellent study base for the correlation of $\chi^{(3)}$ non-linearity and low dimensional structures.

ACKNOWLEDGEMENTS

We thank the SERC and BICC Cables Limited for a CASE studentship award for M. Carroll and the University of Durham for a Special Research Award for this work. We also thank CONACYT-Mexico for an award of a scholarship to R. Rangel-Rojo.

5. REFERENCES

1. C. Sauteret, J.-P. Hermann, R. Frey, F. Pradere, J. Ducuing, R. H. Baughman and R. R. Chance, Phys. Rev. Lett., (1976), 36, 956.
2. E. Hanamura, Phys. Rev. B., (1988), 37, 1273.
3. L. E. Brus, J. Chem. Phys., (1984), 80, 4403.
4. Y. Wang, Acc. Chem. Res., (1991), 24, 133.
5. H. Yokoyama, Science, (1992), 256, 66.
6. Y. Wang, A. Suna, J. McHugh, E. F. Hilinski, P. A. Lucas and R. D. Johnson, J. Chem. Phys., (1990), 92, 6927.
7. C. S. Winter, C. A. S. Hill and A. E. Underhill, Appl. Phys. Lett., (1990), 58, 107.
8. C. S. Winter, S. N. Oliver, R. J. Manning, J. D. Rush, C. A. S. Hill and A. E. Underhill, J. Mater. Chem., (1992), 2, 443.
9. G. H. Cross, M. Carroll, T. L. Axon, D. Bloor, R. Rangel-Rojo, A. K. Kar and B. S. Wherrett, Proc. SPIE, (1992), 1 in press.
10. L. R. Melby, R. J. Harder, W. R. Hertler, W. Mahler, R. E. Benson and W. E. Mochel, J. Am. Chem. Soc., (1962), 84, 3374.
11. A. Rembaum, V. Hadek and S.P.S. Yen, J. Am. Chem. Soc., (1971), 93, 2532.
12. H. Kita and K. Okamoto, J. Appl. Polym. Sci., (1986), 31, 1383.
13. J. Tanaka, M. Tanaka, T. Kawai, T. Takabe and O. Maki, Bull. Chem. Soc. Jap., (1976), 49, 2358.
14. J.B. Torrance, B. A. Scott and F. B. Kaufmann, Sol. St. Commun., (1975), 17, 1369.
15. M. Konno and Y. Saito, Acta Cryst., (1974), B30, 1295.
16. M. Konno, T. Ishii and Y. Saito, Acta Cryst., (1977),B33, 763.
17. I. Shirotani and H. Kobayashi, Bull. Chem. Soc. Jap., (1973), 46, 2595.
18. H. Shinojima, J. Yumoto and N. Uesugi, Appl. Phys. Lett., (1992), 60, 298.

Conjugated Polymer Electro-optic Devices

D. D. C. Bradley,[1*] A. R. Brown,[1] P. L. Burn,[2]
J. H. Burroughes,[1] R. H. Friend,[1] N. C. Greenham,[1]
R. W. Gymer,[1] A. B. Holmes,[2] A. M. Kraft,[2]
and R. N. Marks[1]

[1] DEPARTMENT OF PHYSICS, CAVENDISH LABORATORY, MADINGLEY ROAD, CAMBRIDGE CB3 OHE, UK
[2] UNIVERSITY CHEMICAL LABORATORY, LENSFIELD ROAD, CAMBRIDGE CB2 1EW, UK

1 INTRODUCTION

Conjugated polymers are polymeric semiconductors which combine the desirable processing characteristics inherent of polymer systems with the sought after electrical, electro-optic and nonlinear optical properties of semiconductors[1]. Moreover, being naturally quasi-one-dimensional they have an electronic structure that exhibits features that have to be engineered in the more conventional (three dimensionally bonded) inorganic semiconductors through elaborate growth and patterning procedures that are often costly and time consuming. Electro-optic responses are especially interesting for potential device applications and three distinct types are seen in conjugated polymers. The first, and simplest, is the perturbing effect of a DC electric field which leads to a strongly field dependent refractive index in the vicinity of the dominant π-π^* absorption[2,3]. Since conjugated polymers are in general centrosymmetric this response arises from the DC Kerr effect (the Pockel's coefficient vanishes by symmetry) with the underlying mechanism a quadratic Stark shift of the exciton responsible for the π-π^* absorption[2,3]. The response seen is generally large with a maximum nonlinear susceptibility of order 10^{-8} esu, but the linear absorption in the same spectral range is such that consideration of the appropriately loss-normalised figure of merit indicates that the effect is probably too small to be used in a device[2].

The second type of electro-optic response observed in conjugated polymers is that which accompanies unipolar charge injection in a semiconductor field effect device (e.g. MIS capacitor or MISFET transistor)[4]. Here the electro-optic response is a direct consequence of the low dimensional electronic structure which results in self-localisation of charges added to the polymer chain and formation of polaronic excited states that combine the charge with a lattice distortion[5]. These polaronic states (singly charged polaron or doubly charged bipolaron) have energy levels that lie within the π-π^* optical gap of the pristine material and that are formed by pulling a π and a π^* state into the gap. As a consequence, their formation results in strong oscillator strength shifts which allow modulation of the complex refractive index and offer opportunities for construction of field effect optical modulators and switches[4].

The third electro-optic response, and the one which will be discussed in detail within the rest of this review is that which arises following bipolar charge injection in diode structures. When the two oppositely charged carriers, injected from the two electrodes, can combine to form an exciton and that exciton can subsequently decay radiatively to the ground state, electroluminescence is observed. This light emission in response to an applied bias is the basis of a light emitting diode and as we shall outline below, the observation of this phenomenon in conjugated polymers provides a major stimulus for their continued investigation and development.

Conjugated polymer electroluminescence is a rapidly expanding field [6-28], and there is a strong possibility that these materials will find commercial use in a variety of display applications. Since the first literature report on poly(p-phenylene vinylene) [PPV] (c.f. figure 1) polymer devices[6], following an earlier patent application[7], there have been a rapidly increasing number of publications demonstrating the ubiquity of the phenomenon amongst fluorescent conjugated polymers, with many of the materials anticipated in the patent application now experimentally confirmed as suitable for use in EL devices[9,11,12]. Important and continuing improvements in all features of device performance are being reported, and the colour range accessible already includes the three colours (red, green and blue) required for full colour displays[9,11-13,15,19,24,25]. The combination of polymer processing techniques for low cost film deposition and an emissive light output operating at low DC drive voltages with high brightness and good efficiency is very attractive for large area displays. Moreover, unlike many other sectors of the electronics industry, large area EL panels do not at present have a firmly entrenched material of choice.

The competition faced by conjugated polymers for EL applications falls into four main categories[29], namely (i) inorganic semiconductor light emitting diodes, (ii) fluorescent ion doped inorganic semiconductor thin film devices, (iii) inorganic semiconductor powder devices, and (iv) sublimed molecular dye devices[30,31]. Commercial products are available based on (i), (ii), and (iii) but there are drawbacks with all three, especially for large area applications and there are definite openings for new tehnologies in this area[21].

Figure 1 shows the chemical structures of some of the conjugated polymers that are discussed in this article. We will not discuss here the details of polymer synthesis or film deposition and processing. References are given to earlier papers where these details may be found.

2 CONJUGATED POLYMER ELECTROLUMINESCENT DEVICES

The first reported conjugated polymer device[6] consisted of a very simple sandwich structure which under forward bias showed a turn on at about 25V (threshold field 2 x 10^6 V/cm) with a yellow green emission. The efficiency (in terms of photons emitted per electronic charge transported) was less than or of order 0.01%, a value too small to be of interest for commercial applications but this has subsequently been greatly enhanced, as described below, to values that are now commercially significant. The EL emission spectrum, shown in figure 2, closely resembles the photoluminescence [PL] spectrum of PPV[32], a fact which implies that the same excited state, namely a singlet exciton, gives rise to both emissions. This in turn is consistent with the EL resulting from double charge injection with positive and negative carriers travelling through the polymer film and combining to form an exciton. An estimate of the theoretical limit on the efficiency achievable with PPV devices can be made from considerations of spin statistics. For an exciton formed by non-geminate pair combination (as in EL) there are four possible ways to combine the half-integer spins of the two charge carriers. Of these three give a resultant spin of one, i.e a triplet state, and only one gives a zero-spin singlet state. The triplet state is not linked by a dipole allowed transition to the ground state and thus does not contribute to EL emission. In the case of geminate pair combination (which dominates for PL), the resultant state is always a singlet. We thus expect that the EL yield will be less than or equal to one quarter that observed in PL.

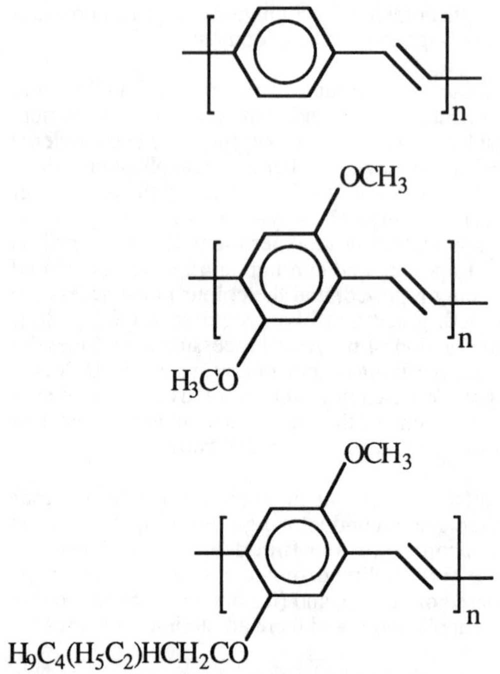

Figure 1 Chemical structures of selected conjugated polymer EL materials: poly(p-phenylenevinylene) [PPV], poly(2,5-dimethoxy-p-phenylenevinylene) [PDMeOPV], and poly(2-methoxy, 5-(2'-ethyl-hexoxy)-p-phenylenevinylene) [MEH-PPV].

This treatment is over simplistic, ignoring as it does both the influence of the energy difference between singlet and triplet excited states on the combination process and also the possibility for interconversion between them. This simple treatment does however provide a useful rough guide to what can be expected. For PPV the PL efficiency is estimated, from the dominant component in the decay which has time constant ≈ 250ps[33], to be as much as 25%. This assumes that the radiative lifetime is ≈ 1ns, a reasonable figure given the known decay times of related oligomeric molecules[34], and the observation of components in the decay with characteristic times of up to this value but not in excess of it[33]. These data suggest therefore that an EL efficiency of up to 6% is a realistic target, a figure that would make PPV highly attractive for commercial applications. The question to be answered is how can such an enhanced value be realised? Other important questions that must be considered are: (i) how to control the device characteristics to allow bright emission at low drive voltages and currents?, (ii) how to tune the colour of emission?, and (iii) how to engineer multi-colour, multi-pixel elements and more complex device structures desired for application purposes? The rest of this article illustrates several of the lines that have been followed in Cambridge to achieve some of these goals. Reference is also made to the work of other research groups active in this field.

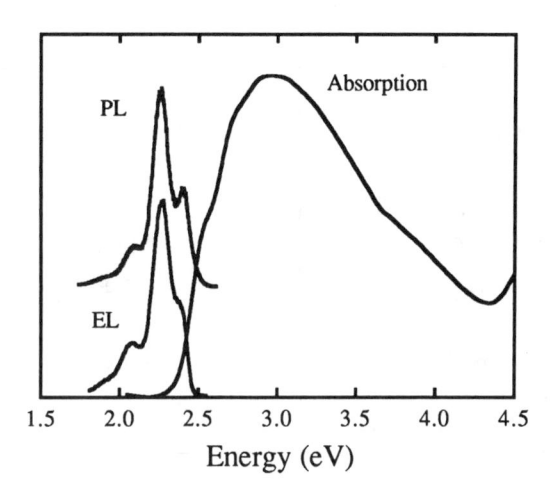

Figure 2 PPV absorption and emission spectra. Photoluminescence [PL] was excited by the 457.9 nm output from a CW Ar+ ion laser. Electroluminescence [EL] is for an ITO:PPV:Ca device. Very similar results are seen for other electrode combinations e.g. ITO:PPV:Al. The different levels of self-absorption are clearly seen in the different lineshapes of the highest energy (0-0) vibronic peak. In the EL spectrum this peak shows up only as a shoulder whereas in the PL spectrum (offset vertically for clarity) it is clearly resolved.

The approaches to modifying the device characteristics fall into two main categories, namely (i) electrode engineering, and (ii) polymer engineering. The former concerns optimisation of the electrodes to match, as far as possible, and within the limits of the available materials, the electronic structure of the polymer. The latter offers scope for molecular level engineering of quantum efficiency, colour, and transport and injection properties, and for supramolecular level engineering both of device structure through lateral patterning, and of transport and injection properties though multilayer device fabrication.

3 DEVICE PHYSICS AND CHARACTERISATION

We consider two processes in the operation of our devices, firstly formation of the exciton and secondly its decay. The former encompasses (i) carrier injection, (ii) carrier transport, and (iii) carrier combination, and the latter involves a competition between radiative and non-radiative decay channels.

Formation of excitons

Carrier injection and transport. Balanced injection and transport are prerequisites for highly efficient operation. If either of these is not achieved then an unproductive (without emission) current is likely to flow and recombination will become localised in a region at, or near to, one of the polymer/electrode interfaces where quenching effects may be important.

Table 1   Examples of suitable electrode materials for conjugated polymer electroluminescence devices

Hole injection (high Φ)	Electron injection (low Φ)
Indium-tin-oxide (ITO)	Al
Al:Al$_2$O$_3$	Mg/Ag
Au	Ca
Pt	In
p-type Si	n-type Si
p-doped polyaniline	etc
etc	

To ensure balanced injection it is necessary to equalise the barriers that arise at the two polymer/electrode interfaces due to the difference in the workfunction of the positive (negative) electrode and the ionisation potential (electron affinity) of the polymer. Moreover, to ensure that injection is readily achieved at low drive voltages we want these barriers to be small. There is unfortunately very little experimental data in the literature on the ionisation potentials and electron affinities of conjugated polymers and the values reported for theoretical calculations show a large spread. This makes it rather difficult to select electrode materials with work functions that match the polymer electronic structure. There is also a large uncertainty in the work function values for the electrode materials, with significant variations due to different deposition conditions and due to composition differences in e.g. the chemical structure of indium-tin-oxide (ITO) coatings.

For the combination of electrode materials used in the first PPV device discussed above the aluminium acts as an electron injecting contact and the aluminium with oxide surface layer as a hole injecting contact. Many other possible combinations of electrode material exist and several examples for both electron and hole injection materials are listed in Table 1. The use of a transparent ITO film as the hole injection electrode allows the EL emission to readily exit from the device, an obvious requirement for applications. Varying the electrode materials used in PPV devices produces results[16] that are consistent with estimates for the expected barriers. It must however, be re-emphasised that the barrier heights are not known accurately. Moreover, little is currently known about the nature and extent of any interfacial layers formed between the polymer film and the electrode material. For PPV prepared via the precursor route, HCl is a by-product of the thermal conversion, a substance widely used to etch ITO. Consequently, even though most of the HCl should escape from the film under conversion in-vacuo, there is still likely to be an interfacial layer at the ITO electrode surface. Similarly, it has recently been shown that e.g. evaporation of Al onto PPV results in part in a chemical reaction of the Al with the polymer, which again may lead to an interfacial layer[35].

The use of a polymer emissive layer lends itself to fabrication of devices on a variety of substrates and these need not be planar or rigid. ITO can readily be deposited on many different substrates, including flexible plastic sheet and curved preforms. Other electrode materials, including e.g. doped conjugated polymer films such as polyaniline[36], can also be used to allow fabrication of flexible diode structures. This diversity in device construction is an additional attractive feature of conjugated polymer EL devices.

To achieve balanced carrier transport requires that the injected positive and negative charge carriers have similar mobilities. The evidence to date is, however, that positive carrier transport is strongly favoured. Thus, for instance, in time-of-flight mobility measurements no negative carrier transport signals are observed[37]. Very low negative carrier mobilities are seen in many organic solids due in part to the strong trapping capability of oxygen[38]. Additional evidence for the imbalance comes from EL spectra for which the amount of self-absorption, observed as an asymmetry in the 0-0 emission line, is significantly stronger than seen in PL spectra (c.f. figure 2). For PL, recombination occurs in regions that extend through the sample but with regions near to the illuminated surface weighted more strongly in accordance with the concept of an absorption depth. The greater self-absorption of EL is consistent with that emission arising from recombination which, on average, takes place deeper within the film i.e. closer to the negative electrode. There is then a long self-absorption path for the emission viewed through the ITO coated glass substrate.

In order to compensate for the imbalance in carrier transport it is possible to fabricate devices that incorporate carrier transport layers whose properties complement those of the emissive polymer[17]. For PPV we require an electron transporting material, of which there are rather few organic examples. Guided by earlier work on heterostructure devices utilising sublimed molecular films[30,31], we have chosen the molecule 2-(4-biphenylyl)-5-(4-*tert*-butylphenyl)-1,3,4-oxadiazole [PBD]. A device was constructed starting with ITO coated glass onto which a 3500Å PPV layer was formed, a 300Å PBD layer was then deposited on top of the PPV in the form of a solid state dispersion in poly(methylmethacrylate) [PMMA]. Finally a Ca negative electrode was evaporated on top of the PBD/PMMA layer. The latter layer has the dual rôle of transporting negative charges away from the negative electrode and also of inhibiting the transport of positive carriers straight through the device. This inhibition of positive carrier leakage currents is expected to have the added benefit of causing the build up of a positive space charge at the PPV:PBD/PMMA interface that will increase the field across the PBD and should give enhanced electron injection (c.f. figure 3). With this electron-conducting, hole-blocking layer the efficiency of the device increased from 0.1% to 0.8% at brightnesses of order 500 Cd/m^2. At lower brightness the efficiency was 1.1%, slightly higher, with the decrease in efficiency with increasing brightness possibly attributable to onset of hole leakage currents, alteration of the PBD/PMMA through sample heating, or exciton quenching in the presence of a high space charge density. Confirmation that electron injection and transport limits device performance comes from experiments in which a hole transporting compound (in the form of a polymer or a dispersion in a polymer matrix) was used in conjunction with a PPV film[26]. In this case the overall efficiency of the device decreased relative to that of an otherwise identical PPV device without the charge transporting layer.

Combination of carriers. The third step in exciton formation involves the combination of oppositely charged carriers. In conjugated polymers these carriers are expected to be polaron-like self-localised excitations[5]. For non-degenerate ground state materials, such as PPV, there are expected to be two carrier types, namely singly charged polarons and doubly charged bipolarons, with the latter favoured at high charge density. The energy levels of polarons and bipolarons lie within the π-π* energy gap, with bipolaron levels more strongly displaced into the gap. This has consequences for exciton formation since the energy of the exciton level is essentially the π-π* gap energy. The question is then whether it is energetically favourable for either two polarons or two bipolarons to combine to form an exciton.

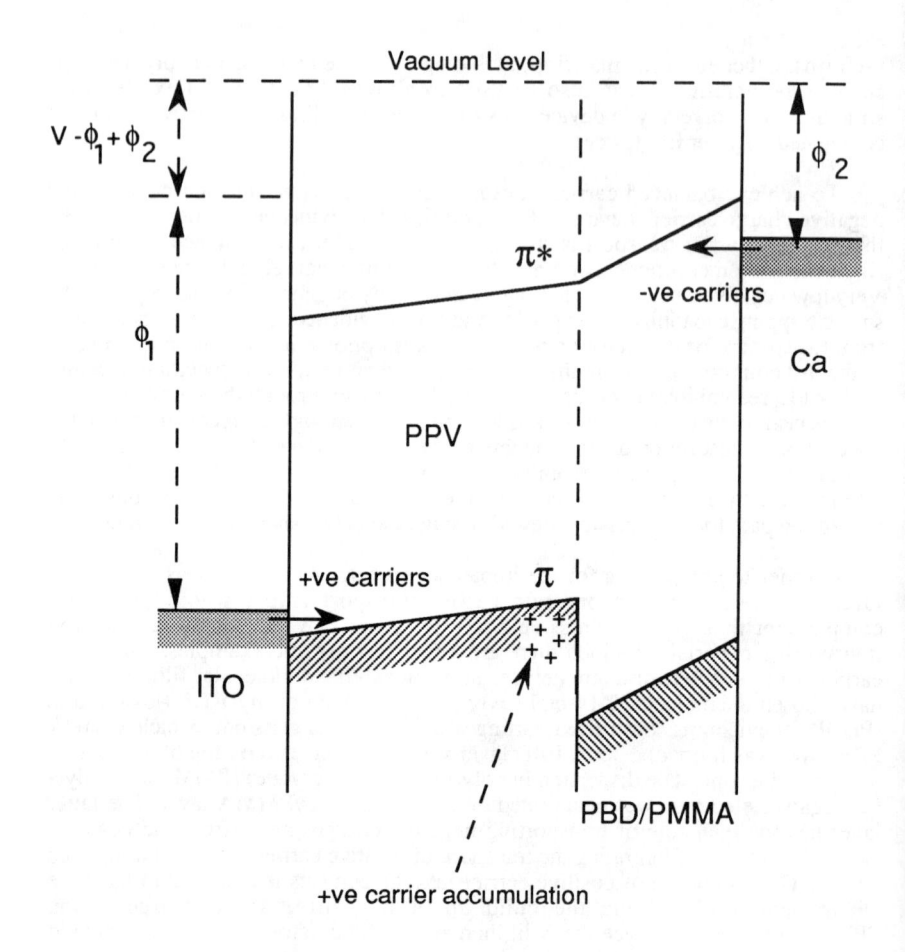

Figure 3 Schematic electronic level structure for ITO:PPV:PBD/PMMA:Ca multilayer heterostructure device under forward bias of V volts. ϕ_1 and ϕ_2 are the workfunctions of ITO and Ca respectively. In the absence of experimental information, the full potential offset between PPV and PBD has for simplicity been drawn at the π electron level. An accumulation of +ve charge at the PPV:PBD/PMMA interface is indicated, and the resulting expected imbalance in the field distribution is illustrated.

No theoretical studies have yet been reported addressing this point. In the case of two polarons the answer is likely to be yes, but the situation of two bipolarons is much less clear. Consideration should also be given to the effects of the high electric fields ($\geq 10^6$ V/cm) present during device operation, upon the extent of lattice relaxation around the injected charges. It is conceivable that lattice relaxation is reduced, making exciton formation more likely.

Experiments have been performed to measure the modulation of transmission through a PPV emissive layer device (ITO:PPV:Ca) in response to a modulated bias field that periodically drives the device into its "on" state (giving light emission)[22]. The measurement uses a double pass geometry for which light is incident through the

ITO contact, through the PPV film and is then reflected back from the metal electrode through the PPV and out through the ITO. This experiment is conceptually similar to that previously reported for modulation of transmission in a field effect device where signatures of polarons and bipolarons were both clearly identified in the modulation spectra[4]. In the case of transmission modulation in a light emitting diode structure we expect additionally to see the absorption signatures of neutral (triplet) excited states. The results bear out this expectation, with a triplet-to-triplet transition readily discernable, just as is seen in photoinduced absorption[32]. Bipolarons are also detected by the presence of two correlated absorption peaks but polarons are not seen, unlike the case of the field effect devices. The bipolarons do not however appear to contribute significantly to the electroluminescence emission since the bipolaron signal saturates for applied voltages in a range where the electroluminescence signal is still increasing strongly. Saturation is possibly suggestive of either there being a limited number of sites at which the bipolarons exist or of an equilibrium between two polarons and a bipolaron that is space charge controlled. The absence of polaron signatures is readily understood if, as seems reasonable, the polarons are either being used up to form singlet and triplet excitons or pass straight through the device. The expected time for this is 100 ns or less (taking a mobility of $10^{-4} cm^2/Vs$, with an applied field of 10^6 V/cm and a film thickness of 100 nm), which is much too short to allow their detection in the mechanically chopped cw experiment that we have performed. That we do see bipolarons, then suggests that they are probably trapped within the film.

Also of consequence to the process of exciton formation are the spin states of the charge carriers that combine. We have already used arguments based on spin statistics to discuss the theoretical limit on EL efficiency. The importance of spin state to the combination processes has been directly observed in optically detected magnetic resonance experiments that detect microwave induced transitions between the Zeeman split levels of the various excitations generated following charge injection[18]. Large electroluminescence, photoluminescence, and conductivity detected magnetic resonance signals have been recorded that support the assignment of a double charge injection process as the mechanism of operation in PPV devices[18]. The results also support the idea that bipolarons are undesirable for efficient emission with conductivity and electroluminescence quenching resonances that can be explained in terms of the field induced enhancement of the conversion of two polarons into a bipolaron. This process quenches electroluminescence by reducing the population of polarons and by simultaneously creating non-radiative recombination centres (see below) and quenches conductivity through the formation of less mobile bipolarons.

Exciton decay

As a consequence of the fact that the same singlet exciton gives rise to both EL and PL we can make use of the many earlier studies of photoluminescence[32] to gain an insight into the processes competing with EL emission. The reason why PL yields in fluorescent conjugated polymers are not as high as in their small molecule organic dye analogues is that the non-radiative decay processes are more efficient. As already commented above the quantum yield for PPV films is estimated to be of order 25% at room temperature. This compares with yields approaching 100% for small molecule analogues in solution, e.g. *bis*-isopropyldistyrylbenzene, effectively a dimer of PPV, has a quantum yield of 94%[34]. What distinguishes fluorescent conjugated polymers like PPV from non-fluorescent ones like poly(2,5-thienylene vinylene)[39] is currently a topic of great interest both from the viewpoint of scientific curiosity and now also from the viewpoint of selecting materials for potential use in EL devices. In general there are many possible non-radiative decay mechanisms that are well known in organic solids. Amongst these are included: (i) multiphonon emission, (ii) intersystem crossing from singlet to triplet manifold, (iii) exciton-exciton fusion at high excitation densities, (iv) migration to quenching sites, (v) excited state energy transfer to even parity singlet states, (vi) self-trapped exciton formation and tunnelling to ground state configuration coordinate surface, etc. Mechanisms (ii), (iii), and (iv) are readily shown to be relevant to PPV and its derivatives by respectively, photoinduced absorption experiments in which there are clear triplet-triplet absorption signals[32], from the intensity dependence of PL decay curves[32], and from a variety of experiments[40] showing quenching of PL by dopant induced, photoexcited and field effect injected, charge carriers.

Mechanism (v) becomes important when the lowest lying excited singlet has the same (even) parity as the ground state, since then by Kasha's rule the transition that should yield fluorescence is dipole forbidden. Recent Pariser-Parr-Pople Quantum Chemical calculations[41] show that a strong alternation in the electron transfer integrals along a polymer chain leads, other than for strong electron correlation effects, to the lowest lying excited singlet having a dipole allowed fluorescent relaxation path to the ground state. P-phenylene moieties within a conjugated chain are predicted to be especially efficacious in achieving this result and it thus appears more than an accident of fate that the materials which have shown most promise for electroluminescence applications do indeed contain such a p-phenylene group. In PPV the lowest lying two-photon accessible excited singlet state has been shown to lie some 0.5 eV above the onset for the lowest single-photon allowed transition[42], an excited state ordering that is in agreement with theory and with the experimentally observed strong fluorescence.

Mechanism (vi) has been discussed as a model to reconcile the variations in fluorescence yield seen for different polydiacetylene polymers and alkyl-substituted polythiophenes, but there is little direct evidence to support the existence of the self-trapped exciton state[43], and the involvement of a low lying two-photon accessible singlet state must be ruled out as an origin of the observed behaviour before this can be taken more seriously.

Spin dependent magnetic resonance measurements can provide extremely useful information on the details of excitation dynamics and the interconversion of singlet and triplet states. Photoinduced absorption detected magnetic resonance measurements have given important confirmation of the nature of excited states in conjugated polymers[44] and a combination of electroluminescence and conductivity detected magnetic resonance measurements on EL diode structures provide a useful means to study in detail the processes involved in device operation. Preliminary results have already revealed several interesting features, and apparently show the unexpected presence of two distinct triplet states[18].

4 COLOUR TUNING AND FILM PATTERNING

Control of the emission colour is clearly a requirement for the achievement of full-colour displays. Emission from fluorescent conjugated polymers is generally observed as a broad band, with superposed vibronic structure, that emerges at energies just below the onset of the π-π* absorption spectrum (c.f. figure 2). The approach to colour tuning is thus to select known examples of fluorescent conjugated polymers or synthesise new ones that have π-π* energy gaps at energies just above those desired for the emission. PPV has a π-π* energy gap of $\approx$ 2.5 eV and its EL emisson is yellow. To tune to the red, one selects polymers with lower π-π* energy gaps. Examples of materials satisfying this criterion include dialkoxy substituted PPV polymers (π-π* energy gaps $\approx$ 2.1 eV) such as poly(2,5-dimethoxy-p-phenylene vinylene) [PDMeOPV][45] (c.f. figure 1) and MEH-PPV[9] (c.f. figure 1), poly(3-alkyl thiophenes) such as poly(3-dodecylthiophene) (π-π* energy gaps ≈ 2.1 eV)[11,24,46], and various copolymers, e.g. of PPV and PDMeOPV (π-π* energy gaps in the range ≈ 2.1 eV to 2.5 eV)[14,16]. Tuning towards the blue involves choosing polymers with larger π-π* energy gaps. Examples of these include poly(p-phenylene) (π-π* energy gap ≈ 2.7 eV)[12], poly(9,9-dihexylfluorene) (π-π* energy gap ≈ 2.9 eV)[13], and copolymers containing interruptions in the conjugated sequences, e.g. thermally converted methoxy leaving group PPV[15,19].

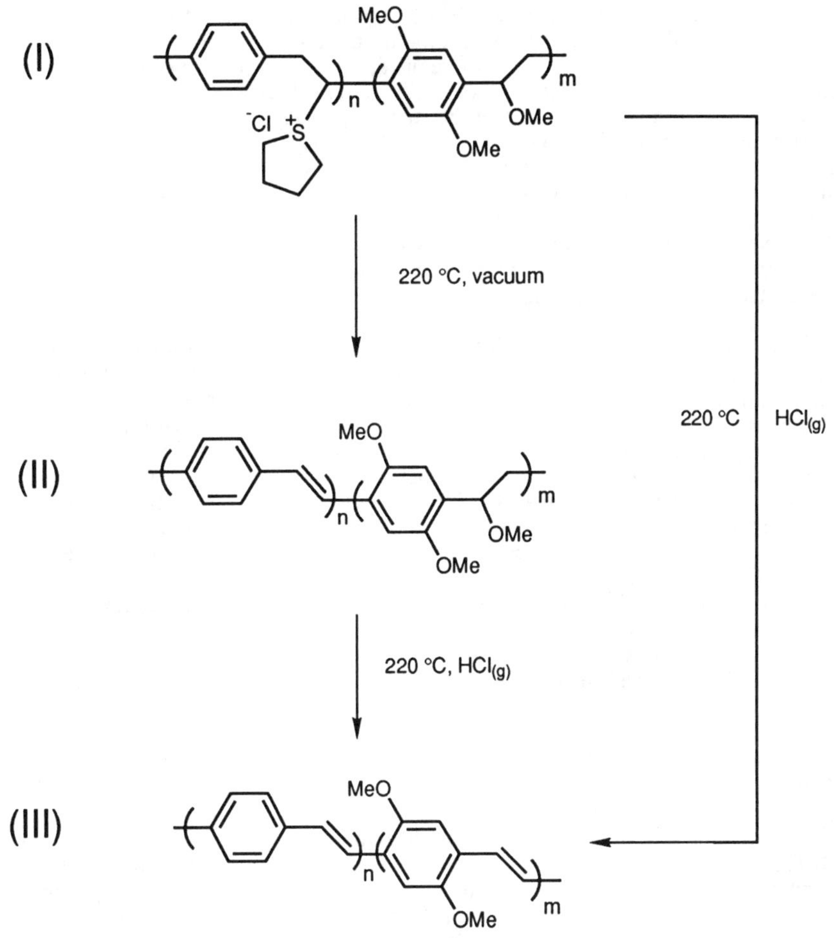

Figure 4 Precursor route synthesis of PPV-PDMeOPV copolymers. The precursor (I) contains two different leaving groups with methoxy moieties preferentially present next to methoxy substituted phenylene rings. Thermal conversion in vacuo yields the conjugated/non-conjugated copolymer (II), which requires subsequent heating under HCl gas to catalyse the elimination of the methoxy moieties in order to generate the fully conjugated polymer (III). Polymer (III) may also be reached direct from precursor (I) by thermal conversion in the presence of HCl gas. N.B. HCl is a by-product of the elimination of the tetrahydrothiophenium moiety, and judicious control of its presence allows control of the compositional balance between polymer structures (II) and (III).

Copolymers offer considerable flexibility for fine-tuning the emission colour. We have studied copolymers of PPV and PDMeOPV and have seen a linear shift in the emission spectra with varying composition[14]. These copolymers in fact allow both red and blue shifts in emission from that seen for PPV. The red shift is as expected following the introduction of segments of PDMeOPV, which as a homopolymer has a lower π-π* energy gap than PPV[45]. The blue shift is not expected and is achieved by careful control of the in-vacuo elimination reaction that converts the precursor polymer into its conjugated form. The synthesis (c.f. figure 4) of the copolymer precursors introduces two different leaving groups with one of these selectively introduced on the PDMeOPV precursor units (polymer (I) in figure 4). The elimination of this methoxy group requires more severe conditions than for elimination of the standard tetrahydrothiophenium [THT] moiety. In particular, acid is required as a catalyst to remove the methoxy groups (yielding polymer (III)) whereas heat alone suffices to eliminate the THT groups. As a consequence, if films are subject only to in-vacuo thermal conversion (from polymer (I) to polymer (II)), the PPV precursor units are preferentially converted with the PDMeOPV units remaining largely unconverted. The result is a polymer in which limited sequences of PPV units are separated from each other by non-conjugated segments (polymer (II)). This conjugated/non-conjugated polymer has a larger π-π* energy gap on account of the limited extent of conjugation, and is thus blue shifted relative to PPV.

These copolymers offer an additional attractive feature in that the control of the conversion reaction allows patterning of a film[14]. To selectively control, within a single film, the areas that undergo full conversion and those that undergo conversion of only the PPV sequences is simply a matter of laying down a patterned capping layer of e.g. aluminium. This can be done by standard photolithographic techniques using a lift-off process. The sample is then thermally treated in vacuo under the standard conditions for elimination of THT leaving groups. The difference here is that the capping layer acts to trap HCl, a by-product of the elimination of THT leaving groups, and this acid then catalyses the elimination of the methoxy leaving groups. Consequently in the capped regions full conversion (to polymer (III)) occurs whilst in the uncapped regions the conversion is limited to PPV sequences (yielding polymer (II)). As a result, patterns can be made in the copolymer film, with the different regions possessing different π-π* energy gaps and different refractive indices. This approach offers the possibility of creating multi-colour elements within a single film and of defining waveguide structures.

Another approach to the attainment of multi-colour emission spectra is to fabricate multilayer devices in which several different polymers, with different π-π* energy gaps are deposited one on top of another[27]. This can be readily achieved with the use of a combination of precursor route (e.g. PPV and copolymers) and/or soluble (e.g. MEH-PPV) conjugated polymers. Sequential deposition by spin-coating is possible when the different materials have complementary solvents (e.g. THT PPV precursor is soluble in methanol but not chloroform whereas MEH-PPV is soluble in chloroform but not methanol). Alternatively, a partial heat treatment following deposition of a precursor polymer layer renders it insoluble before the next layer is spun down. This allows multilayers to be fabricated even when the different polymers have a common solvent in their processable form. Such multilayer devices show emission from regions extending over 200 nm and provide important insights into to the physics of exciton formation[27].

5 ASSESSMENT AND CONCLUSIONS

The extremely rapid progress to date in the development of conjugated polymer electroluminescence bodes well for the future application of these materials. Red, green and blue emission colours have all been demonstrated. For PPV, efficiencies have risen by two orders of magnitude from a paltry 0.01% to a new "best" figure of 2% (albeit at low brightness operation) for a device incorporating an electron transport layer[47]. This figure is already competitive with commercial EL materials. Brightnesses in excess of 1500 Cd/m^2 are now achievable, sufficient for interior display applications, and the lower brightnesses of order 100 Cd/m^2 needed for computer screen displays are readily achieved at moderate current levels. Threshold voltages for light emission of 4 V and less can also be demonstrated. Moreover, these figures are not expected to be the ultimate performance figures for these materials. Improvements in efficiency to values of 5% are anticipated and concomitant improvements in power consumption should follow. The aromatic based precursor-route polymers such as PPV and derivatives appear well suited to fabrication of robust devices and their processing procedures allow large area device fabrication on a range of different substrates and patterning to form multi-colour elements within a single film. It is important to emphasise that lifetime tests performed on the devices that we currently fabricate in a standard laboratory environment are unlikely to give good indications of the lifetime achievable for devices fabricated under rigorous manufacturing (e.g. clean room) conditions. Nevertheless, devices have been run under experimental conditions for periods of up to 3 days without significant degradation in performance. Further purification of the polymers and improved deposition techniques should also assist. The stability is expected to increase steadily as the device performance figures improve since e.g. improved injection and transport should lead to a reduction in the electric fields required for device operation and increased efficiency will result in less heat dissipation.

What needs now to happen is the simultaneous optimisation of all the device parameters. In addition more information is required on the nature of the interfaces between the inorganic electrodes and the organic layers and how these affect device operation, and also how they may be controlled. There is also considerable scope for chemical optimisation of the emissive polymer layer to enhance its fluorescence quantum yield and of the transport layers to provide efficient carrier transport with high carrier mobilities.

Finally, there remain many interesting aspects of the science of these devices to be more fully investigated including the detailed nature of the charge carriers (polarons vs bipolarons and the extent of lattice relaxation), the different processes that allow changes in spin state between the singlet and triplet manifolds, and the reasons behind the very different fluorescent yields seen for various conjugated polymers.

6 ACKNOWLEDGEMENTS

We thank the UK Science and Engineering Research Council and Cambridge Research and Innovation Ltd for their support of this work. ¶ PLB is now at Dyson Perrins Laboratory, South Parks Rd, Oxford, OX1 3QY, UK. # JHB is now at Toshiba Cambridge Research Centre, 260 Cambridge Science Park, Milton Rd, Cambridge CB4 4WE, UK.

7 REFERENCES

1. D.D.C. Bradley, Chem. Br., (1991), 27, 719.
2. O.M. Gelsen, D.D.C. Bradley, H. Murata, N. Takada, T.Tsutsui, and S.Saito, J.Appl.Phys. (1992), 1064.
3. G. Weiser, Phys.Rev.B, (1992), 45, 14076.

4. K.E. Ziemelis, A.T. Hussain, D.D.C. Bradley, R.H. Friend, J. Rühe, and G. Wegner, Phys.Rev.Lett., (1991), 66, 2231 and references therein.
5. A.J. Heeger, S. Kivelson, J.R. Schrieffer, and W.-P. Su, Rev.Mod.Phys., 1988, 60, 781.
6. J.H. Burroughes, D.D.C. Bradley, A.R. Brown, R.N. Marks, K. MacKay, R.H. Friend, P.L. Burn, and A.B. Holmes, Nature (London), 1990, 347, 539.
7. J.H. Burroughes, R.H. Friend and D.D.C. Bradley : British Patent Application N° 8909011.2 (1989) and International Patent Application No. PCT/GB90/00584 (1990).
8. D.D.C. Bradley, A.R. Brown, P.L. Burn, J.H. Burroughes, R.H. Friend, A.B. Holmes, K.D. Mackay, and R.N. Marks, Synthetic Metals, 1991, 43, 3135.
9. D. Braun and A.J. Heeger, Appl. Phys. Lett., 1991, 58, 1982.
10. D.D.C. Bradley, A.R. Brown, P.L. Burn, R.H. Friend, A.B. Holmes, and A. Kraft, "Electronic Properties of Polymers" Springer : Solid State Sciences, 1992, 107, 304.
11. Y. Ohmori, M. Uchida, K. Muro, and K. Yoshino, Solid State Commun., 1991, 80, 605. Y. Ohmori, M. Uchida, K. Muro, and K. Yoshino, Jpn.J.Appl.Phys., 1991, 30, L1938.
12. G. Grem, G. Leditzky, B. Ullrich, and G. Leising, Advanced Materials, 1992, 4, 36.
13. Y. Ohmori, M. Uchida, K. Muro, and K. Yoshino, Jpn.J.Appl.Phys., 1991, 30, L1941.
14. P.L. Burn, A.B. Holmes, A. Kraft, D.D.C. Bradley, A.R. Brown, R.H. Friend and R.W. Gymer, Nature (London), 1992, 356, 47.
15. P.L. Burn, A.B. Holmes, A. Kraft, D.D.C. Bradley, A.R. Brown, and R.H. Friend, J.Chem.Soc.Chem.Commun., 1992, 32.
16. P.L. Burn, A. Kraft, A.B. Holmes, A.R. Brown, D.D.C. Bradley, and R.H. Friend, Proceedings MRS Meeting, Boston, December 1991.
17. A.R. Brown, D.D.C. Bradley, P.L. Burn, J.H. Burroughes, R.H. Friend, N.C. Greenham, A.B. Holmes, and A.M. Kraft, Appl.Phys.Lett., 1992, at press.
18. L.S. Swanson, J. Shinar, A.R. Brown, D.D.C. Bradley, R.H. Friend, P.L. Burn, A. Kraft, and A.B. Holmes, Phys.Rev.B., 1992, at press.
19. A.R. Brown, P.L. Burn, D.D.C. Bradley, R.H. Friend, A.M. Kraft, and A.B. Holmes, Proceedings UPS'91, Okazaki Japan, Mol.Cryst.Liq.Cryst., 1992, 216, 111.
20. S. Karg, W. Riess, V. Dyakonov, and M. Schwoerer, Proceedings EMRS Spring Meeting, Strasbourg 1992, Synthetic Metals, 1992, at press.
21. D.D.C. Bradley, Proceedings EMRS Spring Meeting, Strasbourg 1992, Synthetic Metals, 1992, at press.
22. A.R. Brown, K. Pichler, N.C. Greenham, D.D.C. Bradley, R.H. Friend, P.L. Burn, and A.B. Holmes, Proceedings ICSM'92, Gothenburg, Sweden, August 1992, Synthetic Metals, 1992, at press.
23. R.N. Marks, D.D.C. Bradley, R.W. Jackson, P.L. Burn, and A.B. Holmes, Proceedings ICSM'92, Gothenburg, Sweden, August 1992, Synthetic Metals, 1992, at press.
24. N.C. Greenham, A.R. Brown, D.D.C. Bradley, and R.H. Friend, Proceedings ICSM'92, Gothenburg, Sweden, August 1992, Synthetic Metals, 1992, at press.
25. A.B. Holmes, D.D.C. Bradley, A.R. Brown, P.L. Burn, J.H. Burroughes, R.H. Friend, N.C. Greenham, R.W. Gymer, D.A. Halliday, R.W. Jackson, A. Kraft, J.H.F. Martens, K. Pichler, and I.D.W. Samuel, Proceedings ICSM'92, Gothenburg, Sweden, August 1992, Synthetic Metals, 1992, at press.
26. A. Kraft, P.L. Burn, A.B. Holmes, D.D.C. Bradley, R.H. Friend, and J.H.F.

Martens, Proceedings ICSM'92, Gothenburg, Sweden, August 1992, Synthetic Metals, 1992, at press.
27. A.R. Brown, N.C. Greenham, J.H. Burroughes, D.D.C. Bradley, R.H. Friend, P.L. Burn, A. Kraft, and A.B. Holmes, Chem.Phys.Lett., 1992, at press.
28. Proceedings ICSM'92, Gothenburg, Sweden, August 1992, Synthetic Metals, 1992, at press.
29. J.W. Allen : Proceedings International Conference on Electrolumionescence 1992.
30. C.W. Tang and S.A. VanSlyke, Appl.Phys.Lett., 1987, 51, 913. C.W. Tang, and S.A. VanSlyke, J.Appl.Phys., 1989, 65, 3610.
31. C. Adachi, T. Tsutsui, and S. Saito, Appl.Phys.Lett., 1989, 55, 1489. C. Adachi, T. Tsutsui, and S. Saito, Appl.Phys.Lett., 1990, 56, 799. C. Adachi, T. Tsutsui, and S. Saito, Appl.Phys.Lett., 1990, 57, 531.
32. N.F. Colaneri, D.D.C. Bradley, R.H. Friend, P.L. Burn, A.B. Holmes, and C.W. Spangler, Phys. Rev. B, 1990, 42, 11670.
33. I.D.W. Samuel, B. Crystall, G. Rumbles, P.L. Burn, A.B. Holmes, and R.H. Friend, Proceedings EMRS Spring Meeting, Strasbourg 1992, Synthetic Metals, 1992, at press.
34. I.B. Berlman, Handbook of Fluorescence Spectra of Aromatic Molecules, Academic Press, New York and London, 1971.
35. W.R. Salaneck : private communication.
36. G. Gustafsson, Y. Cao, G.M. Treacy, F. Klavetter, N. Colaneri, and A.J. Heeger : Nature (London), 1992, 357, 477.
37. T. Takiguchi, D.H. Park, H. Ueno, and K. Yoshino, Synth. Metals, 1987, 17, 657. J. Obrzut, M.J. Obrzut, and F.E. Karasz, Synth. Metals, 1989, 29, E103. P. Strohriegl and D. Haarer, Makromol.Chem.Macromol.Symp., 1991, 44, 85.
38. M. Stolka, Proceedings EMRS Spring Meeting, Strasbourg 1992, Synthetic Metals, 1992, at press.
39. A.J. Brassett, N.F. Colaneri, D.D.C. Bradley, R.A. Lawrence, R.H. Friend, H. Murata, S. Tokito, T. Tsutsui, and S. Saito, Phys.Rev.B, 1990, 41, 10586.
40. D.D.C. Bradley and R.H. Friend, J.Phys.Condens. Matter, 1989, 1, 3671. H.B. Gu, T. Takiguchi, S. Hayashi, K. Kaneto, and K. Yoshino, J.Phys.Soc.Japan, 1987, 54, 3997. K.E. Ziemelis, A.T. Hussain, D.D.C. Bradley, R.H. Friend, J. Rühe, and G. Wegner, Phys.Rev.Lett., 1991, 66, 2231.
41. Z.G. Soos, S. Etemad, D.S. Galvao, and S. Ramasesha, Chem.Phys.Lett., 1992, 194, 341.
42. C.J. Baker, O.M. Gelsen, and D.D.C. Bradley, Chem.Phys.Lett., 1992, at press.
43. T. Kobayashi, Proceedings EMRS Spring Meeting, Strasbourg 1992, Synthetic Metals, 1992, at press.
44. Z.V. Vardeny, Proceedings EMRS Spring Meeting, Strasbourg 1992, Synthetic Metals, 1992, at press.
45. H.S.Woo, S.C. Graham, D.A. Halliday, D.D.C. Bradley, R.H. Friend, P.L. Burn, and A.B. Holmes, Phys.Rev.B, 1992, 46, 7379.
46. J. Rühe, N.F. Colaneri, D.D.C. Bradley, R.H. Friend, and G. Wegner, J.Phys: Condens. Matter, 1990, 2, 5465.
47. N.C. Greenham, unpublished.

The Design and Characterization of Polymers for Second-order Nonlinear Optics

David J. Williams

CORPORATE RESEARCH LABORATORIES, EASTMAN KODAK COMPANY, ROCHESTER, NY 14650-2110, USA

1 INTRODUCTION

The field of second-order nonlinear optics in polymers has progressed considerably in recent years. Progress in the field has been documented in recent publications.[1,2] The interest and progress in this field have been spurred by the need for relatively inexpensive materials technology for active optical and integrated optical devices. Polymeric materials composed of chromophores with large molecular hyperpolarizability that are aligned in a polar fashion exhibit large, useful second-order nonlinear optical coefficients. Due to the flexibility of polymer design and synthesis many of the required solid-state properties required for fabricating thin films for nonlinear optics can be optimized simultaneously.

One of the most useful classes of chromophores that have been incorporated into polymeric structures are the aromatic sulfones (1).

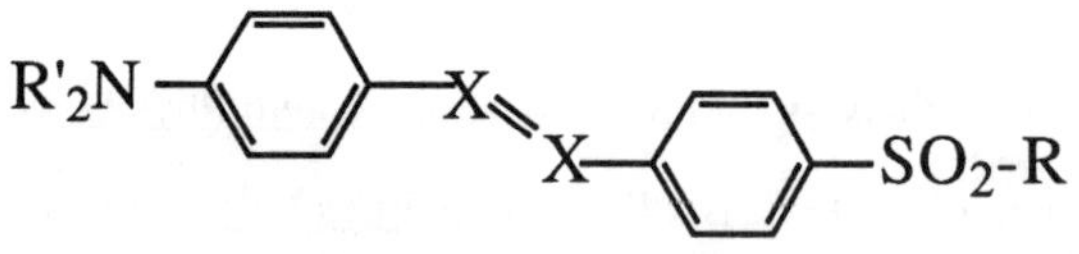

R'_2N—⟨⟩—$X=X$—⟨⟩—SO_2-R 1a X = C

1b X = N

Their utility derives from two factors. The first is the ability to functionalize the molecule from the acceptor end. This enables considerable flexibility of polymer design. For instance, both side- and main-chain structures[3] can be synthesized where the polar axis of the molecule can be perpendicular or parallel to the chain contour. These architectural differences might be expected to exhibit different electric field alignment and relaxation characteristics. Functionalization of the sulfone group can also be used to optimize polymers for incorporation into Langmuir-Blodgett films for subsequent fabrication of noncentrosymmetric film structures.[4] A second advantage is the ability to electronically modify the acceptor strength by perfluorination of methylene units α to the sulfone group.[5] This increases the acceptor strength similar to that of a nitro group without as large a red shift in the optical absorption spectrum.

In this chapter recent progress in our laboratory in the design and characterization of this class of polymers will be discussed. The contrast in orientation and relaxation behavior of side- and main-chain systems will be focused upon. In addition the design, fabrication, and unique properties of sulfone polymers optimized for Langmuir-Blodgett multilayers will also be discussed.

2 SIDE-CHAIN VERSUS MAIN-CHAIN SYSTEMS

In this section various issues pertaining to side-chain and main-chain polymer systems are discussed. In the first part of the section differences in orientation and relaxation behavior of a selected side-chain and main-chain polymer are discussed. In the second part the issue of cooperativity between dipolar sections in main-chain systems is discussed.

<u>Orientation and Relaxation</u>

A comparative study of poling and relaxation behavior was conducted on the following main- and side-chain polymer systems.[6]

The synthesis and characterization was reported previously.[3] Inspection reveals a fundamental difference between the side-chain (**2**) and main-chain (**3**) systems having to do with the orientation of the dipole moment and hyperpolarizability tensor. In polymer **2** these vectors are relatively uncorrelated with respect to the backbone contour whereas in the MC polymers these vectors are highly correlated with the backbone contour. Because of this the poling process is expected to have a considerably different impact on overall polymer orientation in these two systems. In **2**, relatively little reorientation would be expected. In the main-chain system the high degree of correlation between the molecular dipole direction and the chain contour would be expected to cause an alteration of the end-to-end vector of the polymer during the poling process. The impact of this on the poling and relaxation process is discussed below.

An experiment was devised to probe the relationship between poling induced $\chi^{(2)}$ and the orientation polarization induced by the poling process.[3] It involves the simultaneous measurement of the second-harmonic signal intensity (SHG) and the thermally stimulated discharge current (TSD) in the poled film. In the experiment a sample is heated near its glass temperature. An electric field is applied and orientational polarization is induced. The temperature is lowered to ambient and the field is then removed. Subsequently, the voltage source is replaced by an ammeter and the current is measured as the sample is heated while measuring the second-harmonic signal. The current consists of contributions from the depolarization of dipole orientation, ionic displacements, and space charge from injection at the electrodes. The temperature dependence of the second-harmonic signal will be primarily associated with dipole orientation. The relationship between the SHG and TSD signals is given by

$$[(I_{2\omega}(T)/I_{\omega}(T_1))]^{1/2} = (1 - 1/P)\int_{T_1}^{T} (j(T')/q)\,dT' \qquad (1)$$

where $I_{2\omega}(T)$ and $I_{\omega}(T_1)$ are the second-harmonic intensities measured during the heating process and at the initial temperature, respectively; P is the orientation polarization; $j(T')$ is the contribution to the TSD current from the orientation depolarization process; and q is the heating rate. The advantage of measuring the TSD signal is that it has a considerably higher signal-to-noise ratio than the SHG signal. Since the TSD signal bears a first derivative relationship to the SHG signal a more accurate determination can be made of the temperature at which the rate of decay of orientational depolarization is a maximum. The simultaneous observation enables an unambiguous assignment of the TSD signal due to orientational relaxation.

The results of the experiment described above obtained for **2** under two sets of experimental conditions are shown in Figures 1a and b.

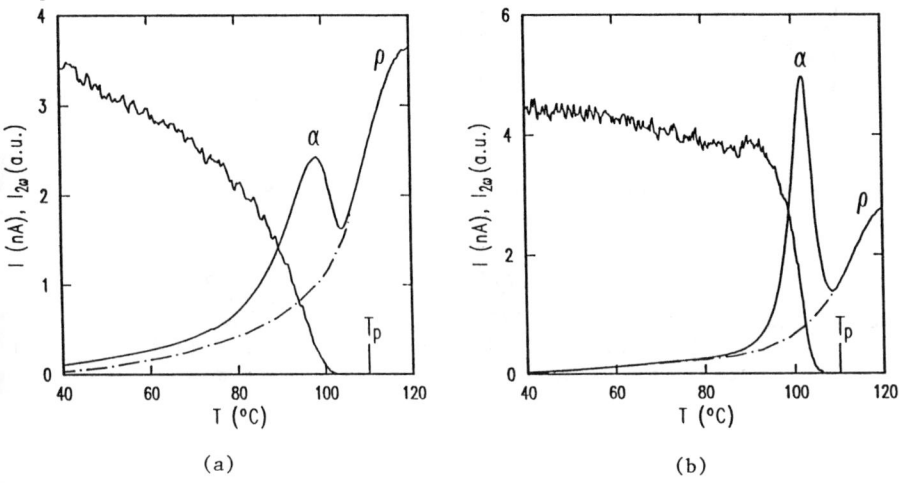

FIGURE 1. Combined TSD/SHG experiment on the side-chain polymer **2**: (a) Poled at 110 °C, 30 V, 30 min. (b) Poled at 110 °C, 30 V, 10 min, and cooled at 3 °C/h.

In the first experiment the sample was quenched rapidly to room temperature following poling at 110 °C; approximately 10 °C above its T_g. The two peaks in the TSD trace are labeled α and ρ. Comparison with the SHG data clearly indicates that the SHG signal and α contribution to the TSD traces are correlated. The sign of the ρ peak is consistent with recombination of displaced ionic charge. The experiment in Figure 1b was conducted in a similar fashion except that the sample was cooled at 3 °C/h. The α signal is considerably sharper and much of the relaxation occurring at lower temperature has been eliminated. The dramatic effect is attributed to physical aging of the polymer, which eliminates excess free volume. Thus, a relatively simple processing step makes the orientationally induced polar order considerably more stable in time.

A similar experiment (Figure 2a) was performed on polymer **3**.

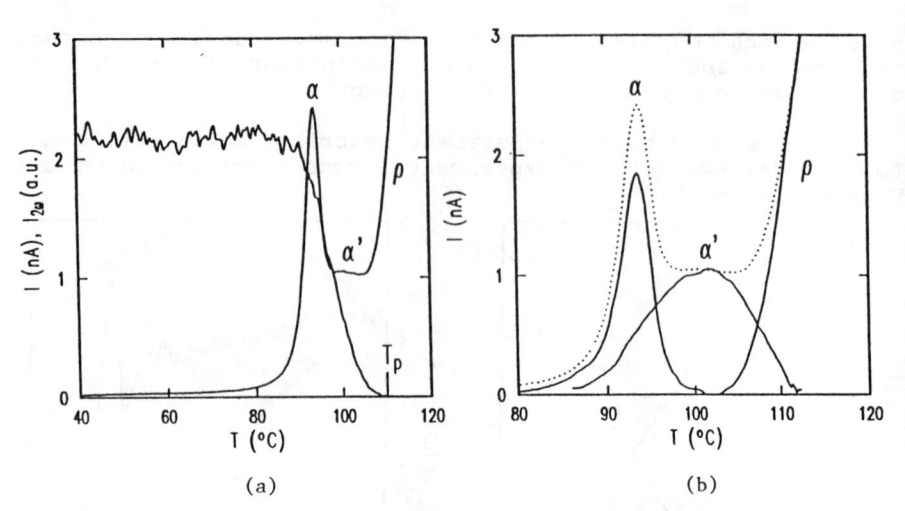

FIGURE 2. Combined TSD/SHG experiment on the main-chain polymer **2**: (a) Poled at 110 °C, 30 V, 30 min, and cooled at 3 °C/h. (b) Deconvolution of the TSD spectrum into α, α', and ρ peaks.

Here the sample was poled at 110 °C, which was about 18 °C above its T_g. Similarly, thermal annealing was shown to virtually eliminate thermally induced relaxation until the sample reached the vicinity of T_g. A distinct difference was observed relative to the side-chain polymer. An additional contribution to the TSD was observed and termed α'. Careful examination of the curves shows that the maximum rate of decay of the SHG coincided with the position of the α' peak. Differentiation of the SHG signal produced the α' peak shown in Figure 2b. By adjusting only the relative amplitudes of the α and ρ contributions the experimental TSD curve could be reproduced by adding these contributions together. Clearly the side- and main-chain systems behave differently.

An explanation for the observed behavior can be made by referring to Figure 3. The rectangles represent the plane of the aromatic system tangent to the chain contour. Molecular orbital calculations indicate that the dipole moment is tipped by 37° relative to the plane of the ring. Components of the dipole moment can be defined in the molecular plane and perpendicular to it. From INDO and MOPAC calculations we know that β, the molecular hyperpolarizability, is essentially parallel to the ring.

The experimental data can be understood if we assume that the ring system has two relaxation components, one occurring at T_g and the other occurring at the temperature of the α' peak (in this case about 18 °C above T_g). The lower temperature component is attributed to a relaxation motion

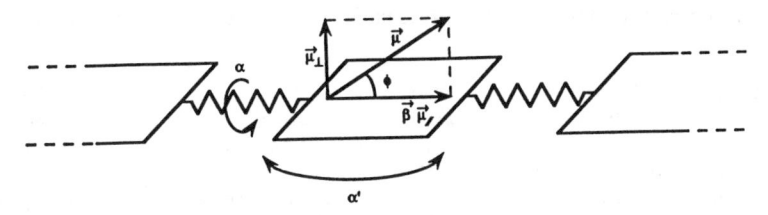

FIGURE 3. Schematic drawing of the rigid chromophore and the flexible spacer units in the main-chain polymer.

around the axis that is tangent to the chain contour. Thus the component of the macroscopic polarization associated with perpendicular components of the microscopic dipole moment are responsible for the α relaxation. Since the largest component of the β tensor is parallel to the chain direction little SHG signal is lost. At the α' relaxation motions of the chain contours randomize the contributions from the parallel dipolar components as well as the component of β, which is responsible for the SHG signal. Analysis of the calculated dipole moment components indicate that the ratio of the α' to the α components of the polarization should be 2.04. Integration of the respective peaks in the TSD curves gives 453 nC and 253 nC, respectively, or a ratio of 1.75. Considering the level of approximation there is very good agreement between this model and the experimental data.

Another consequence of this model is that reorientation of the end-to-end vector of the polymer chain must accompany the induction and relaxation of poling-induced order. This is evident from considerations of the geometrical relationship of the molecular dipole moment relative to the chain contour and from the fact that temperatures well above the T_g of the polymer are required for the relaxation to occur. This architectural feature may very well be a key to enhancing the stability of poling-induced order in these types of systems.

Cooperativity

One of the potentially intriguing features of polar main-chain polymers is the possibility of cooperativity leading to enhanced values of $\chi^{(2)}$ in these systems. From a qualitative perspective, one could argue that if dipolar units in a main-chain polymer are linked together, they should no longer behave as independent units. Since the poling efficiency is proportional to the dipole moment this could lead to a higher degree of orientation in the poling field. Upon closer examination the chemical nature of the linking units becomes very important. Bond geometries and motional restrictions will determine whether cooperativity is observed.

In the polymer 3 no cooperativity was observed. The electric field-induced $\chi^{(2)}$ was about what one would expect based on the dipole moment and the vector component of the hyperpolarizability tensor (β) value of the repeat unit. This can be understood from statistical considerations of polymer chains and data on several model systems.

If we assume that the dipole moment (μ_i) and the vector component of the hyperpolarizability tensor (β_i) are analogous to the bond vectors (l_i) linking the units of the polymer chain then quantities for the polymer chain dipole moment (M) and chain vector component of the hyperpolarizability (B), which are similar to the end-to-end vector of a polymer chain (r), can be defined. For an extended rigid chain M and B should approach the product of the degree of polymerization (n) with μ and β, respectively. In this case a large enhancement of $\chi^{(2)}$ relative to that for an equivalent concentration of monomeric species would be expected.

Assuming that rotation of the units about the chain is unrestricted (an unrealistic assumption) and otherwise has fixed valance angles θ, the quantities M and B can be defined as follows:

$$M = \sum_{i=1}^{n} \mu_i \qquad (2)$$

$$B = \sum_{i=1}^{n} \beta_i \qquad (3)$$

M and B are simply vector sums over the individual monomeric units. Average squares of M and B may also be written as

$$<M^2> = \sum_{i=1}^{n}\sum_{j=1}^{n} \mu_i \cdot \mu_j \qquad (4)$$

and

$$<B^2> = \sum_{i=1}^{n}\sum_{j=1}^{n} \mu_i\, \mu_j \qquad (5)$$

The solutions to the summations in equations 4 and 5 was discussed by Flory[7] and are

$$<M^2> = n\mu^2(1 - \cos\theta)/(1 + \cos\theta) \qquad (6)$$

and

$$<B^2> = n\beta^2(1 - \cos\theta)/(1 + \cos\theta) \tag{7}$$

The expression for the electric field-induced $\chi^{(2)}$ in the case where $\mu E \ll kt$ is given by[8]

$$\chi^{(2)}_m = N_m f_m E\mu\beta/kt \tag{8}$$

where N_m is the number density of nonlinear chromophores, and f_m is a collection of local field factors. Using the relationships defined above, an expression for the susceptibility of a polymer chain can be written as

$$\chi^{(2)}_p = (N_p n f_p E\mu\beta/kt)(1 - \cos\theta)/(1 + \cos\theta) \tag{9}$$

where f_p is the collection of local field factors for the polymer chain. If experimental conditions are chosen so that $N_p n = N_m$, and assuming that $f_m \sim f_p$, an enhancement factor (G) can be written as

$$G = \chi^{(2)}_m / \chi^{(2)}_p \sim (1 - \cos\theta)/(1 + \cos\theta) \tag{10}$$

If the polymer chain conformation is such that the vector component of β is parallel to the polymer chain and bonding angles connecting it to adjacent segments is close to the typical tetrahedral angle of 109.5°, an enhancement of about 2 relative to an equivalent concentration of uncorrelated monomer units is expected.

Molecular modeling and experiments were conducted on the model structure **4**

4a n = 1; 4b n = 2; 4c n = 3.

Molecular modeling studies showed that the sulfone substituent on the piperidine ring has a C-S-C bond angle of 99°, which is very close to perpendicular. Under these circumstances the value of G subject to the assumptions stated above would be expected to be approximately 1.35.

While dimers and trimers would not be expected to exhibit the conformational behavior of polymeric chains the effect of the C-S-C bond angle should be very apparent in measurements of $\mu\beta$ on these systems, and this is shown in Table 1.[9] Within experimental error no substantial enhancement is seen in the dimer and the results on the trimer compared to the momomeric compound 5 confirm this analysis.

It is therefore not surprising that no enhancement in $\chi^{(2)}$ is seen for the polymer 3 where the C-S-C bond angle would be expected to be similar to that of the model compounds and would therefore dominate the chain statistics of the end-to-end vector of the polymer chain.

Clearly a much more extended bond angle than is permitted by the incorporation of the chromophore into the polymer chain is needed for significant enhancement to occur

3 LANGMUIR-BLODGETT POLYMERS

An intriguing prospect for enhancing second-order nonlinear optical properties is to use the Langmuir-Blodgett technique to deposit polymer films with high order parameters using the hydrophobicity and hydrophilicity differences that can be built into polymer chains. Poling uncorrelated polymer films at reasonable values of electric fields typically yields order parameters in the range 0.1 to 0.15. With Langmuir-Blodgett deposition much higher order parameters are possible.

Penner et al.[4] have shown that noncentrosymmetry can be achieved by y-type LB deposition using alternative layers of an NLO active polymer with an inert spacer polymer. Films were prepared from polymer 5 with polyisobutyl methacrylate or poly-t-butyl methacrylate as the alternating spacer.

Table 1   Product Of Hyperpolarizablity and Dipole Moment for Several Compounds

Compound	$\mu\beta$ (x 10^{48} esu)	$\mu\beta$/monomer unit (x 10^{48} esu)
4a	45	45
4b	104	52
4c	136	45

5

Very high quality multilayer films could be prepared with waveguide dimensions. A 131 bilayer film was prepared that exhibited a waveguide attenuation for 623.8 nm light on the order of 1 db/cm. Measurements of polarized SHG intensity in the reflection mode as a function of input polarization showed that an order parameter $<\cos^3\theta>$ on the order of 0.74 could be achieved in these films. From the hyperpolarizability of the chromophore and these numbers a value for $\chi^{(2)}$ of 1×10^{-7} esu was obtained for these films.

Initial results on these films provided considerable motivation to explore the possibility of phase-matched second-harmonic generation in these films. Experiments were recently conducted by Clays et al.[10] In this study Cerenkov type phase matching was observed with 815 nm fundamental light producing 415 nm harmonic light on a 144 bilayer, 0.504 μm thick film at 830 nm and 1.06 μm as well as 632.8 nm. Loss measurements on this film showed that losses were on the order of 0.5 db/cm at 632.8 nm. Measurements of the dispersion of the film were used to predict Cerenkov radiation output angles. At 430 nm the output coupling angle is 6.1°.

These preliminary results demonstrate the potential utility of LB films for waveguide nonlinear optics. The next stage in this effort is to demonstrate phase-matched SHG between guided modes in the active film. With the high second-order coefficients in these films, very efficient second-harmonic conversion is expected.

REFERENCES

1. P. N. Prasad and D. J. Williams, 'Introduction to Nonlinear Optical Effects in Polymers', Wiley & Sons, New York, 1991.
2. 'Organic Materials for Nonlinear Optics II', R. N. Hahn and D. Bloor eds., The Royal Society of Chemistry, Cambridge, 1992.
3. W. Kohler, D. R. Robello, C. S. Willand, and D. J. Williams, Macromolecules, 1991, **24**, 4589.
4. T. L. Penner, N. J. Armstrong, C. S. Willand, J. S. Schildkraut, and D. R. Robello, Proc. Soc. Photo-Opt. Instrum. Eng., 1991, **1560**, 386.
5. D. R. Robello, A. Ulman, E. J. Urankar, and C. S. Willand, U. S. Patent 5,008,043, 1991.
6. W. Kohler, D. R. Robello, P. T. Dao, C. S. Willand, and D. J. Williams, J. Chem. Phys., 1990, **93**, 9157.
7. P. J. Flory, 'Principles of Polymer Chemistry', Cornell University Press, Ithaca, 1953, p. 399-413.
8. G. R. Meredith, J. G. VanDusen, and D. J. Williams, Macromolecules, 1982, 1385.
9. M. A. Mitchell, M. Tomida, J. K. Hall Jr., H. Hampsh, D. R. Robello, C. S. Willand, and D. J. Williams, to be published.
10. K. Clays, T. L. Penner, N. J. Armstrong, and D. R. Robello, Proc. Soc. Photo-Opt. Instrumen. Eng., 1992, 1775, accepted for Publication.

Mutated Bacteriorhodopsins – New Biopolymers for Optical Information Processing

N. Hampp

UNIVERSITY OF MUNICH, SOPHIENSTRASSE 11, D-8000 MÜNCHEN 2, GERMANY

1 INTRODUCTION

The photochromic retinal protein bacteriorhodopsin (BR) which is related to the human visual pigment is contained in the purple membrane (PM) fragments of the cell membrane from *Halobacterium halobium* as a two-dimensional crystalline lattice. About twenty years after its discovery[1] it is now one of the best investigated membrane proteins[2]. The biological function of BR in the halobacterial cell is that of a light-driven proton pump which generates a proton gradient over the cell membrane under illumination and thereby converts light energy into chemical energy. The energy stored in the proton gradient is utilized by a membrane bound ATP-synthase for ATP generation. The attractive photophysical properties of this biological photochrome, *i.e.* its efficient photochemistry and its excellent stability against chemical, thermal and photochemical degradation is not affected by the isolation of the PM patches from the halobacterial cell. Many proposals appeared for possible technical applications of BR[3,4]. Attempts were made to use the proton-pumping and photoelectric properties of BR for the desalination of sea water[5], for the conversion of sunlight into electricity[6] and for the development of "biochips" with BR as a molecular switch[7]. The optical properties of BR, *i.e.* the extreme stability against radiation, the ps-timescale of the primary photochemical step[8] and its photochromic properties attracted the interest of other groups. New impulses to the field of technical applications were given by the development of a halobacterial transformation system which is the key to the engineering of BR. These techniques allow modification of the photophysical properties of wildtype BR (BR_{WT}) by the exchange of functionally important amino acid residues. Once these new materials have been generated they can be produced in virtually unlimited amounts by conventional biotechnological methods. The use of mutated BRs in holographic applications like real-time pattern recognition[9] is reported in this paper.

2 PHOTOCHEMISTRY AND PHOTOPHYSICS OF BR

The proton translocation from the inner to the outer side of the halobacterial cell membrane is coupled to the photochemical conversions of BR[10] (Fig. 1). The chromophoric group in BR is formed by a retinal molecule which is covalently bound to the protein moiety via a Schiff base linkage to the amino acid lysine-216 and an inner shell of amino acids. In Fig. 1 the different intermediates are given by the commonly used single letter code. Indices indicate the absorption maxima. Thermal conversions are represented by thin black arrows, photochemical transitions by thick grey ones. Intermediates shown in upright letters contain all-*trans*, those in italic letters 13-*cis* retinal. Filled letters symbolize that the Schiff base is protonated, outlined letters (M-states) indicate that the Schiff base is deprotonated.

In the dark a mixed chromophore configuration of all-*trans* (D-state) and 13-*cis* (B-state) is found, so-called dark-adapted BR. Under illumination light-adaptation takes place which results in a pure all-*trans* population (B-state).

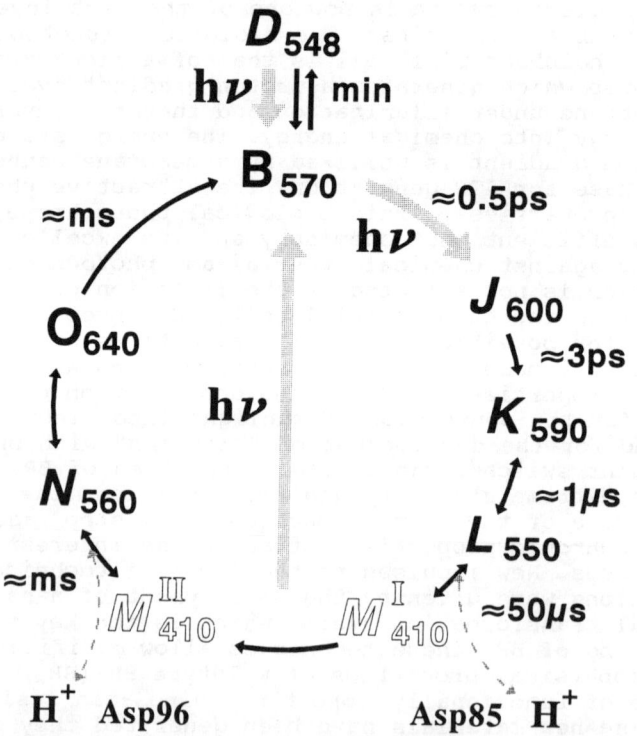

Figure 1 Scheme of the photochemical and thermal conversions of bacteriorhodopsin.

The formation of the photo-intermediate J occurs within a picosecond after absorption of a photon. From there BR cycles through a sequence of thermal relaxations to the M^I state in about 50 μsec and a proton is released from the Schiff base to aspartic acid in position 85 (Asp85). The M^I to M^{II} transition is an essentially irreversible step[10]. Since the M-states are approximately 160 nm blue-shifted from the B-state they dominate the photochromic properties of BR. During the M → N transition the reprotonation of the Schiff base occurs. This step is catalysed by the internal proton donor aspartic acid 96 (Asp96).[11] The reprotonation of Asp96 from the outer medium is accompanied by a configuration change of the retinal chromophore from 13-*cis* to all-*trans* during the N → O transition.

Besides the thermal relaxation from M → B, which takes approximately 10 ms for wildtype bacteriorhodopsin BR_{WT} in aqueous suspension of pH 7, a photochemical pathway from M → B exists which is initiated by the absorption of a blue photon by the M-intermediates.

In a BR variant where the lifetime of the M-intermediates is prolonged by several orders of magnitude, e.g. up to 20 s, almost pure photocontrolled switching of BR between the B- and M-state by yellow (B → M) and blue light (M → B) can be achieved. A material which fulfills this demand is BR_{D96N}. By genetic manipulation an asparagine residue (Asn) was introduced in position 96 replacing Asp96. This leads to a complete loss of the internal proton donor capabilities of BR[12] and shifts the control of the M-lifetime to the extramolecular pH. Therefore changes in the bulk pH allow adjustment of the M-lifetime of BR_{D96N} over several orders of magnitude[13].

Due to the excellent stability of BR in PM-form towards photochemical degradation the state of absorption can be changed reversibly at least several hundred thousand times[14].

The high specific absorption of the B-state of 63,000 l $mol^{-1}cm^{-1}$ and the high quantum efficiency of $\phi = 0.64$[15] of the primary photoreaction B → J result in a high light sensitivity and a strong intensity-dependence of the light-induced changes of the absorption and of the refractive index of a BR-film.

3 HOLOGRAPHIC PROPERTIES OF BR-FILMS CONTAINING BACTERIORHODOPSIN AND ITS MUTATED VARIANTS

Films obtaining PM can be obtained, e.g., by embedding PM into inert matrices like polyvinylalcohol or polyacrylamide. These films can be used as reversible holographic recording media.

For hologram formation and erasure both photochemical conversions B → M ("B-type" holograms) and M → B ("M-type" holograms) can be employed[16]. In relation to the incident intensity pattern a spatially modulated population distribution between the initial B-state and the M-intermediates is induced. In B-type recording the information is written with a wavelength inside the absorption band of the B-state (e.g. 568 nm). M-type recording uses a pump beam which initiates the B → M reaction (e.g. 568 nm) in order to achieve a high initial population of the M-state but the information is recorded with blue light (e.g. 413 nm). The pump beam can be used simultaneously as a readout beam and is not destructive but constructive for the hologram formation. Since both formation and erasure of the holograms are photochemically controlled in M-type recording, this hologram type is used for real-time processing.

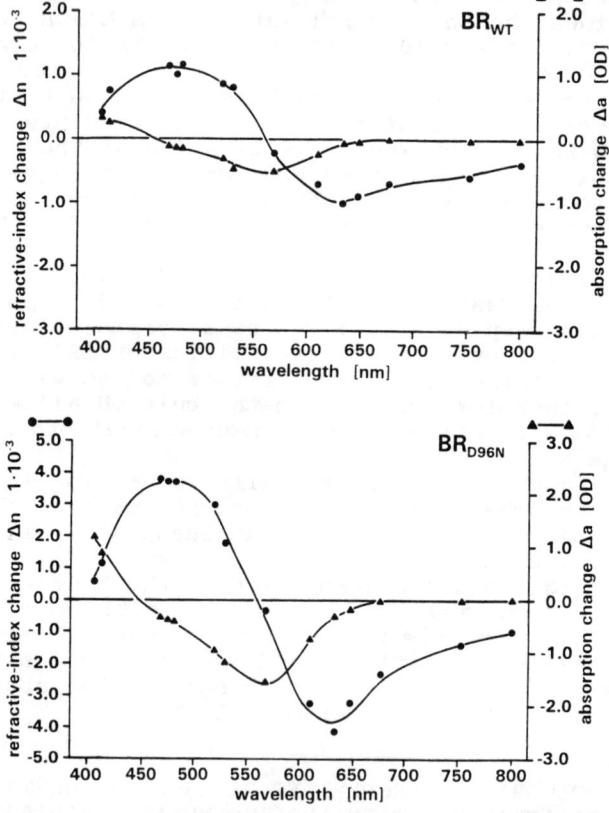

Figure 2 Light-induced changes of the absorption [▲-▲] and the refractive index [●-●] in BR$_{WT}$ and BR$_{D96N}$ films.

A modified Michelson-interferometer, operating in 'pump-probe' mode, is suitable for the characterization of the basic photochemical properties of BR-films, in particular to measure the light-induced changes of the refractive index and the absorption coefficient[17]. The spectral relation of both parameters in BR_{WT} and BR_{D96N} films which have the same optical density of $OD_{570} = 4$ are compared in Fig. 2. The changes in absorption (triangles) and refractive index (dots) under cw-excitation with 20 mW/cm^2 at 568 nm are substantially larger for BR_{D96N} than for BR_{WT}. Light-induced absorption changes of $\Delta\alpha_{max} = 1.5$ are related to refractive-index changes of $\Delta n = 0.004$ at the wavelength 633nm.

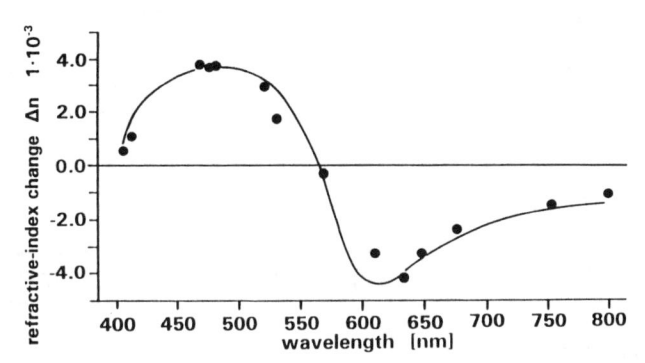

Figure 3 Comparison of the measured (dots) and the calculated (line) values for the refractive index changes in a BR_{D96N}-film.

From the measured light-induced absorption changes the spectral dependence of the refractive index changes in BR-films were calculated by means of the Kramers-Kronig relation. In Fig. 3 the result for the BR_{D96N}-film from Fig. 2 is shown. It can be derived that the retinal containing BR-chromophore can be treated in a first approximation as an undisturbed chromophore. Therefore the Krames-Kronig relation can be applied for the calculation of the refractive index changes on the base of the easily accessible absorption data. Both parameters influence the holographic diffraction efficiency.

The fact that the Kramers-Kronig relation is an acceptable approximation for the spectral dependence of the refractive index changes indicates that the chromophore group in BR is shielded by the protein and interactions between different chromophores are minimized.

Due to the increased modulation of absorption and refractive index in BR_{D96N} (see Fig. 2) an about 50 % higher recording sensitivity and a two-fold higher diffraction efficiency compared to BR_{WT}-films is observed.

The spatial resolution of BR-films is > 5000 mm^{-1} independent on the BR-variant.

A systematic experimental investigation of the diffraction efficiency of BR-films containing BR_{WT}, BR_{D85E} and BR_{D96N} with respect to its intensity-, pH- and temperature-dependencies showed that a balanced choice of several interfering parameters must be chosen to optimize the holographic properties of BR-films[18].

4 APPLICATIONS OF BR-FILMS IN HOLOGRAPHY

The described BR-films have been successfully tested in a number of holographic applications. Examples from polarization Fourier holography, where the photoinducable anisotropy of BR-films is employed to improve the signal-to-noise (S/N) ratio and their application in pattern recognition are presented.

Fourier holograms and polarization holography

In Fig. 4 the optical setup used for Fourier holography is shown schematically. The beam emitted by laser 1 is split into two coherent beams by a polarizing beamsplitter PBS. The beam in the upper arm is expanded and spatially filtered (BE + SF) and then projected onto the BR-film by the Fourier transform lens FTL.

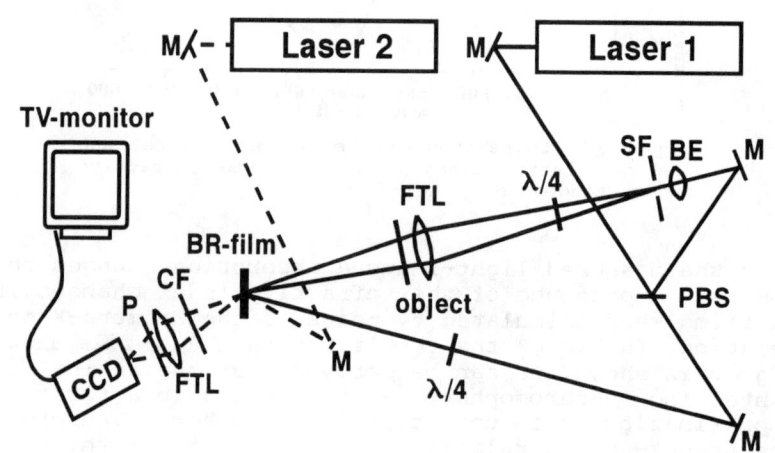

Figure 4 Experimental setup for polarization recording in Fourier holography.

   In this arrangement a post-lens Fourier transformation of the object which is placed directly behind FTL is obtained in the plane of the BR-film[19]. The Fourier pattern of the object is superimposed with the reference beam from the lower arm. The polarization states of the beams in the upper and lower arm can be independently controlled and adjusted, e.g. by two λ/4-plates. The holographic grating formed is recorded in the BR-film and read out by a second beam (laser 2) which has usually a different wavelength. The wave diffracted by the hologram and the transmitted recording beams are separated by means of a color filter CF. The hologram wave is re-transformed onto a CCD array. The photographs in Fig. 5 show reconstructed Fourier holograms of a USAF test pattern, which were taken directly from the TV-monitor connected to the CCD without any further processing. The apparently high S/N-ratio in

Fig. 5a was obtained by employing the polarization properties of BR-films[20]. For means of comparison in Fig. 5b the same test pattern is shown but the polarization filtering was not used.

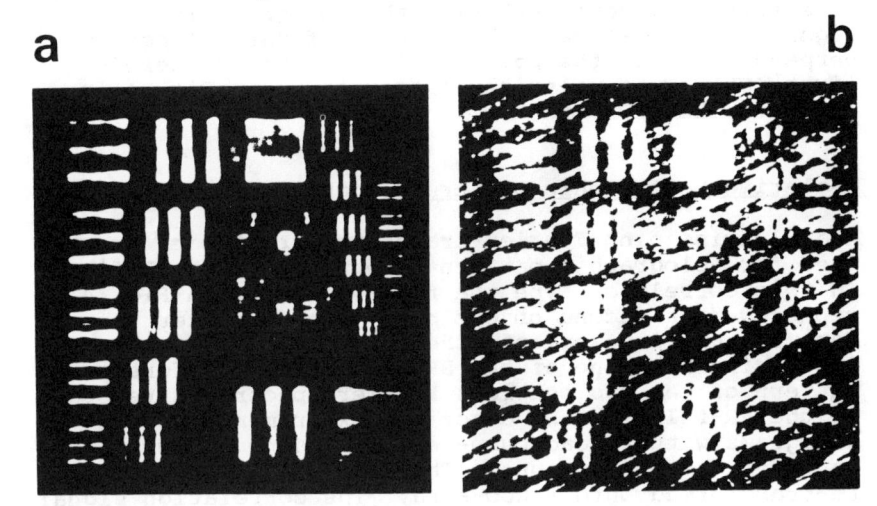

a
b

Figure 5 Screen photograph of a Fourier hologram (USAF test pattern) recorded in a BR-film (a) with and (b) without polarization filtering.

PW1	PW2	PR	PH	SL	RDE
↑	↑	↑	↑	↑	100%
↑	→	↑	→	↑	30%
○	○	↑	⇋	↑	90%
○	○	→	◊	→	90%
○	○	↑	↑	↑	80%
○	○	→	→	→	80%

Figure 6 Polarization recording in BR-films: Relations of the polarization states for B-type and M-type holograms.

Indicated in Fig. 6 is the state of polarization of the two writing beams (PW1 and PW2), the reading beam (PR), the diffracted holographic wave (PH) and the scattered light (SL). The last column gives the relative diffraction efficiency (RDE) of each configuration normalized to that in line 1. In particular the configurations in line 3 and 4 are important in practice. For oppositely circular polarized writing beams a diffracted hologram wave with extreme elliptical polarization is obtained for readout waves with linear polarization. With a linear polarizer the scattered light can be suppressed effectively without weakening the hologram signal significantly, since the long axis of the ellipse is perpendicular to the polarization of the scattered light (SL). This polarizer (P) is shown in Fig. 4 in front of the CCD detector.

Real-time holographic pattern recognition

A dual-axis joint-Fourier-transform correlator[21] was used for holographic pattern recognition with BR-films. In this correlator type the reference beam in Fig. 4 is replaced by a second object beam, whose Fourier transform is superimposed onto the Fourier transform of object 1 in the plane of the BR-film. The interference pattern of the two coherent Fourier spectra is recorded in the BR-film. In the described experiments M-type holograms with a read/pump-beam of 530 nm and writing beams of 413 nm were used. The diffracted wave is re-transformed onto a CCD-array. The correlation signal is monitored with a conventional TV-set. No non-optical post-processing is applied.

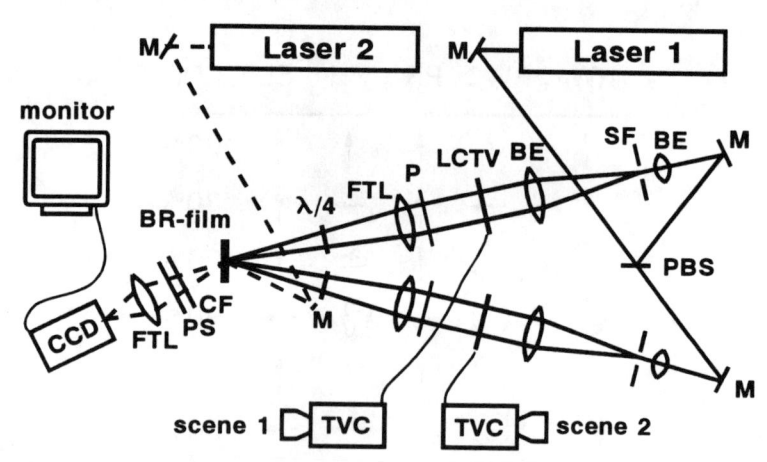

<u>Figure 7</u> Dual-axis joint-Fourier-transform correlator for real-time pattern recognition with BR-films.

As spatial light modulators two miniature liquid crystal TV screens are used (LCTVs). Two TV-cameras catch the scenes and transmit them as standard video signals to the correlator.

In Fig. 8 the result of a holographic correlation of two constantly displayed patterns is shown. An engraving of Munich from the 18th century was used as a 'master' pattern (top). Below the screen photograph of the correlation signal is shown, *i.e.* the bright dot in the middle of the picture, which was obtained with the 'search' pattern shown in its upper right hand corner (bottom).

<u>Figure 8</u>   Holographic pattern recognition with BR-films.

In the case the 'search' criterion changes its position, the related correlation spot moves, too. This is shown in Fig. 9. On one of the LCTVs in Fig. 7 the picture of a sailing boat was displayed constantly (search criterion). In front of the TV-camera connected to the LCTV in the other arm a second picture of the boat was placed and a correlation spot appeared. Moving the boat around resulted in a proportional movement of the correlation spot[9]. Fig. 9 was obtained from a long-term exposure (4 seconds) of a video tape where the tracking of the movement of the boat by the correlator was recorded. The thickness of the line indicates the speed of the boat at any moment.

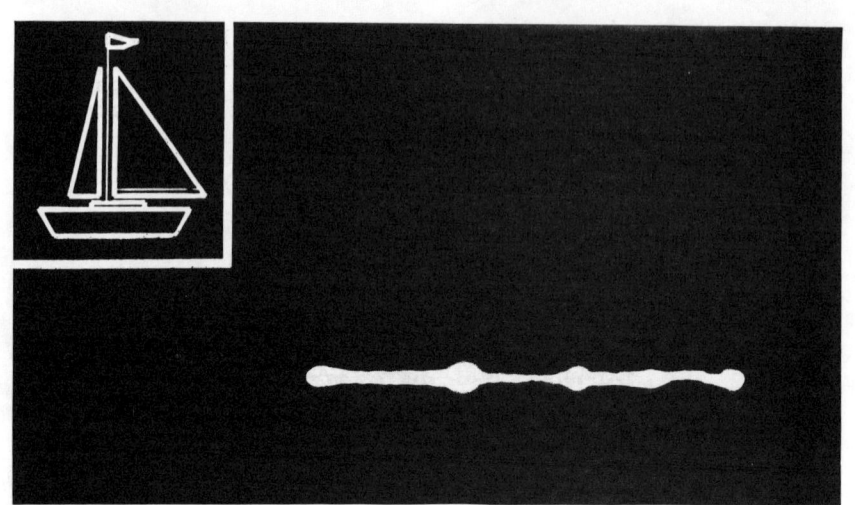

Figure 9 Tracking of the movement of the boat with the real-time correlator.

BR-films for holography containing the BR-variant BR_{D96N} have advantageous properties compared to BR_{WT}-films. In the last years these BR-films have been well characterized and tested in a number of applications[3]. The increasing knowledge of the molecular mechanism of BR and the advanced tools for the modification of this retinal protein by genetic engineering, will in future enable the design of "custom-tailored" BR-variants with specifically optimized optical properties for various applications.

REFERENCES

1. D. Oesterhelt and W. Stoeckenius, Nature (Lond.), 1971, 233, 149.

2. P. Kouyama, K. Kinositu and A. Ikegami, Adv. Biophys., 1988, 24, 123.

3. D. Oesterhelt, C. Bräuchle and N. Hampp, Quart. Rev. Biophys., 1991, 24, 425.

4. N. Hampp, R. Thoma, C. Bräuchle, F.-H. Kreuzer, R. Maurer and D. Oesterhelt, 'Molecular Electronics - Science and Technology', ed. A. Aviram, AIP Conf. Proc., 1992, 262, 181.

5. D. Oesterhelt, FEBS Lett., 1988, 64, 20.

6. K. Singh and S.R. Caplan, TIBS, 1980, 5, 62.

7. F.T. Hong, Biosystems, 1986, 19, 223.

8. M.C. Nuss, W. Zinth, W. Kaiser, E. Kölling and D. Oesterhelt, Chem. Phys. Lett., 1985, 117, 1.

9. R. Thoma and N. Hampp, Opt. Lett., 1992, 17, 1158.

10. J.K. Lanyi and G. Váró, Biochemistry, 1990, 29, 2241.

11. H.J. Butt, K. Fendler, E. Bamberg, J. Tittor and D. Oesterhelt, EMBO J., 1989, 8, 1657.

12. M.S. Braiman, T. Mogi, T. Marti, L. Stern, H.G. Khorana and K.J. Rothschild, Biochemistry, 1988, 27, 8516.

13. A. Miller and D. Oesterhelt, Biochim. Biophys. Acta, 1990, 1020, 57.

14. N. Hampp and C. Bräuchle, 'Photochromism Molecules and Systems', eds. H. Dürr and H. Bouas-Laurent, Elsevier, Amsterdam, 1990, 954.

15. J. Tittor and D. Oesterhelt, FEBS Lett., 1990, 263, 269.

16. N. Hampp, C. Bräuchle and D. Oesterhelt, Biophys. J., 1990, 58, 83.

17. D. Zeisel and N. Hampp, J. Phys. Chem., (in press).

18. N. Hampp, A. Popp, C. Bräuchle and D. Oesterhelt, J. Phys. Chem. 1992, 96, 4679.

19. J.W. Goodman, 'Introduction to Fourier Optics', McGraw-Hill, New York, 1968.

20. N. Hampp, R. Thoma, C. Bräuchle and D. Oesterhelt, Appl. Opt., 1992, 31, 1834.

21. T.C. Lee, J. Rebholz and P. Tamura, Opt. Lett., 1979, 4, 121.

The Design and Development of Photochromic Systems for Commercial Applications

H. G. Heller,[*] C. C. Elliot, K. Koh, S. Al-Shihry, and J. Whittall

SCHOOL OF CHEMISTRY AND APPLIED CHEMISTRY, UNIVERSITY OF WALES, CARDIFF CFI 3TB, UK

INTRODUCTION

Photochromic materials for optical recording on a write/read/erase optical disc require the following combination of properties: (1) complete thermal stability at ambient temperatures, (2) a high conversion of the near colourless form into the more highly coloured form on irradiation with ultraviolet light, (3) high quantum efficiencies for bleaching when optical recording is carried out by visible light on the precoloured form, and (4) fatigue resistance appropriate to the application.

If the optical disc is to be playable on standard equipment which uses a laser diode emitting infrared radiation then the photochromic material must be infrared active.

Non destructive read-out can be achieved by: (1) using a photochromic system in which the coloured form absorbs in the infrared region but the quantum efficiency for bleaching in this region is very low, (2) by taking advantage of a large difference in refractive index between the two forms of the photochromic system, and (3) having an optically active photochromic system in which the read mode utlises the difference in optical rotation of polarised infrared radiation between the two forms.

PREVIOUSLY REPORTED PHOTOCHROMIC FULGIDES

Before describing new developments, the influence of molecular tailoring on the photochromic properties of fulgides containing heteroaromatic rings will be reviewed.

In the fulgides (1), changing the hetero atom (X) from oxygen to sulphur to nitrogen causes the colour of the photocyclised form (2) to change from red to purple to blue respectively.

From the spectrum of fulgide (1,X=O) and its cyclised form (2,X=O) (figure 1), it can be seen that an important feature is that the coloured form has minimum absorption in the near ultraviolet region and thus does not act as an internal filter for the activating radiation[1].

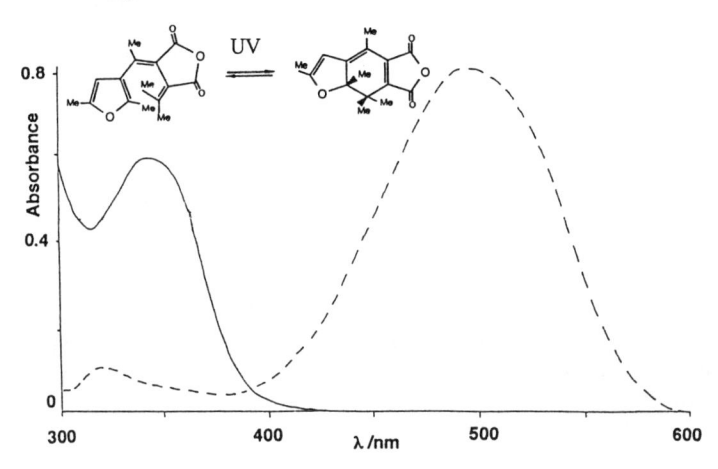

Figure 1 Absorption spectra of fulgide (1,X=O) in toluene (1×10^{-4}M) before(———) and (- - -) after UV irradiation.

Replacement of the methyl group at the 5 position of the furyl group by a phenyl ring containing an electron releasing substituent in the para position, causes a major bathochromic shift in the long wavelength absorption band of the cyclised form and a large hyperchromic effect. Thus the cyclised form with a para N-pyrrolidinophenyl group in place of a methyl group has a molar extinction coefficient of 26260 dm^{3}mol^{-1}cm^{-1} at the wavelength maximum of 590nm compared to 8200 dm^{3}mol^{-1}cm^{-1} at 496nm for the cyclised form of the methyl substituted version[2].

(3)   (4)

Fulgide (3, R=Me) is photochromic and the quantum efficiency for colouring is 20%. However, when R is hydrogen, the photochromic properties are completely lost. Kurita and coworkers[3] found that the quantum yield for colouring increased with the size of the group R. Thus, when R is isopropyl, the quantum yield for colouring is increased to 62%.

In fulgide (5), replacement of the isopropylidene group by the bulky inflexible adamantylidene group causes the 7,7a-sigma bond of the cyclised form (6) to weaken with a consequent improvement in the <u>quantum efficiency for bleaching</u> from 5.5% to 30%.[4] Replacement of the methyl groups by cyclopropyl groups causes a bathochromic shift in the long wavelength absorption band of the fulgide due to the double bond character of cyclopropyl groups[5].

Fulgide (7) (cf. figure 2) which has both isopropyl and adamantylidene groups, has a <u>quantum efficiency for colouring</u> at 366nm of 50% and a <u>quantum efficiency for bleaching</u> at 546nm of 26% in toluene solutions[6].

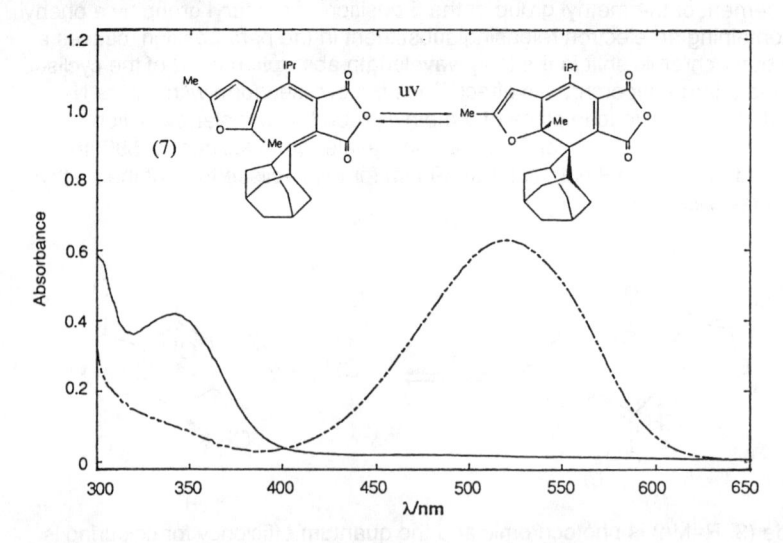

Figure 2 Absorption spectra for open and cyclised forms of fulgide (7) in toluene (1x10^{-4}M).

Fulgimides [eg (8)] show photochromic properties similar to those of the corresponding fulgides as shown in figure 3. They have comparable resistance to photodegradation.

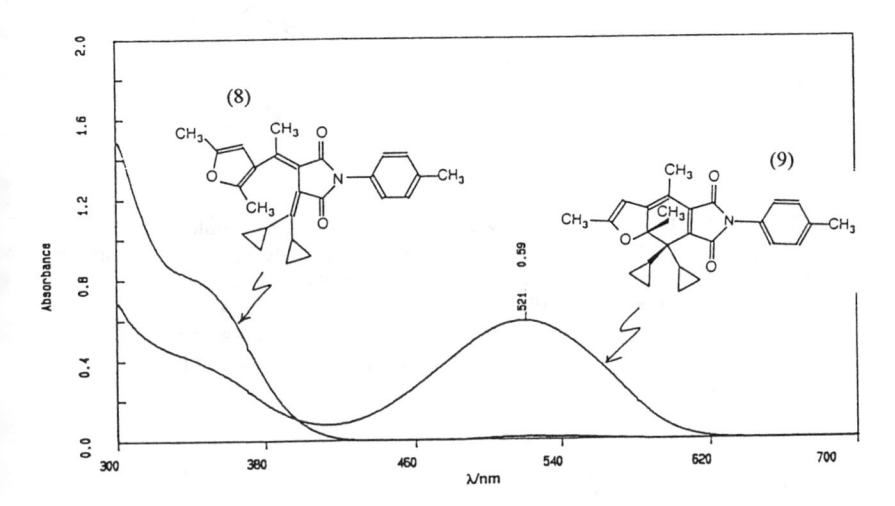

Figure 3 Absorption spectra of open and cyclised forms of fulgimide (8) in toluene (1×10^{-4}M)

Photochromic Langmuir - Blodgett films [7] and liquid crystals [8] based on fulgimides have been reported.

PHOTOCHROMIC ISOFULGIMIDES

Having introduced the name 'fulgimides' to describe the photochromic imides based on fulgides[9], the name 'isofulgimides' is now introduced for the isomers of fulgimides in which one or other of the carbonyl groups of the anhydride ring of fulgides is replaced by a substituted nitrogen. The alpha isofulgimide (10) is defined as the one which has the carbonyl group as part of the conjugated system in its corresponding cyclised form (11), while the beta isofulgimide (12) has the doubly bonded nitrogen as part of the conjugated system in its cyclised form (13).

Alpha isofulgimides

Beta isofulgimides

The isofulgimides can be prepared by cyclisation of the appropriate succinamic acid with dicyclohexyl carbodiimide (DCC). Treatment of the succinamic acid (16) or the isofulgimide (18) with acetyl chloride results in their cyclisation or rearrangement to the corresponding fulgimide (17). The succinamic acid derivative (16) can be obtained by reaction of fulgide (14) with a primary aliphatic or aromatic amine or by the reaction of the appropriate succinic half ester with the Grignard derivative of the amine.

The reaction of phenylhydrazine with fulgide (14) can give either the beta isofulgimide or fulgimide derivatives (20) and (21) respectively, depending on the cyclisation conditions.

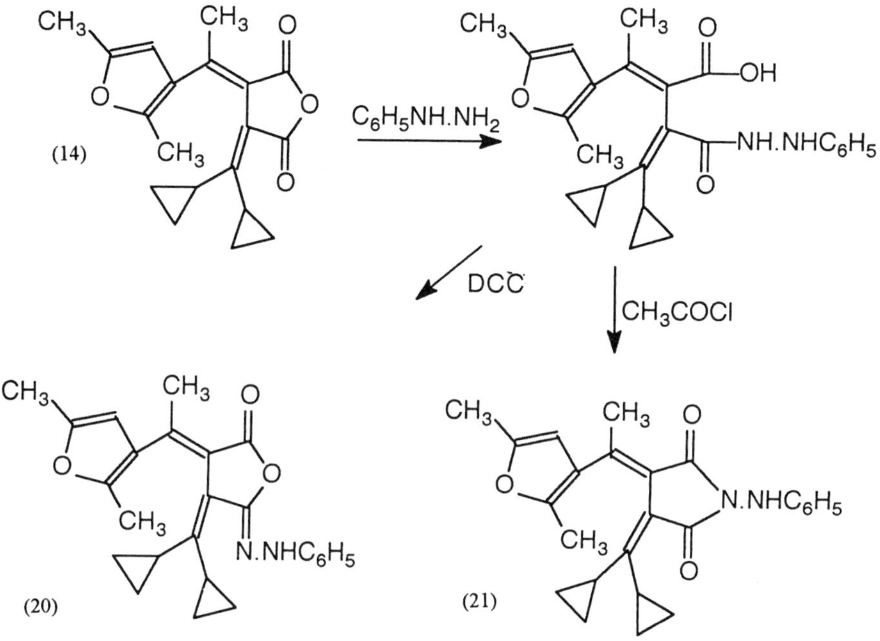

The scheme shown below exemplifies the reactions of succinic half esters (22) and (23) with the Grignard derivative of N-phenyl-N-methylhydrazine to give alpha and beta isofulgimides (24) and (25) respectively.

Fulgide (14) and the corresponding beta isofulgimide (24) show similar photochromic properties with reversible colour changes from pale yellow to red. The main difference is that the molar extinction coefficient of the long wavelength absorption band of the cyclised form from the beta isofulgimide (24) is about 2.5 times greater than that of the cyclised form of the fulgide (14) as shown in figure 4.

The cyclised form of the alpha isofulgimide (25) shows a large hypsochromic shift of its long wavelength absorption band (λ_{max} 438nm) compared to the cyclised form of the fulgide (14) (λ_{max} 500nm) of the beta isofulgimide (24) (λ_{max} 497nm) and of the fulgimide (26) (λ_{max} 527nm) as can be seen from figures 4-6.

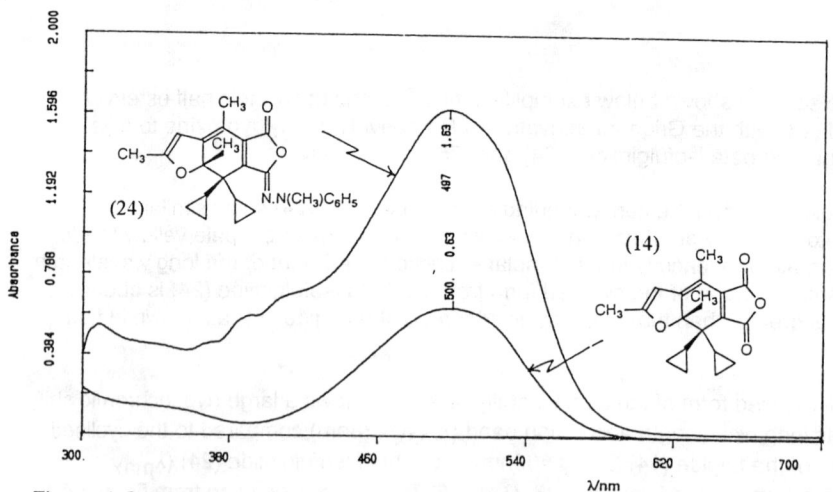

Figure 4 Comparison of the absorption spectra of cyclised fulgide (14) with cyclised beta isofulgimide (24) in toluene (1×10^{-4} M)

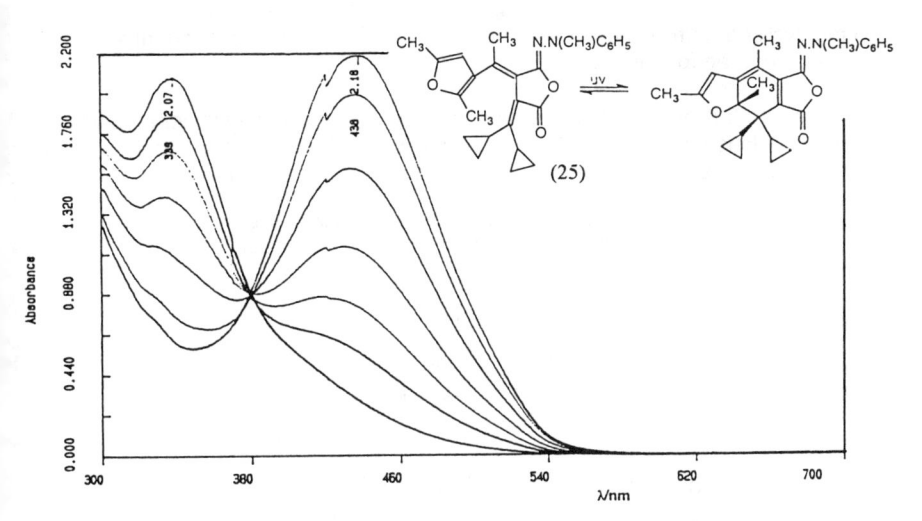

Figure 5 Photochromism of alpha isofulgimide (25) in toluene (1x10^{-4}M)

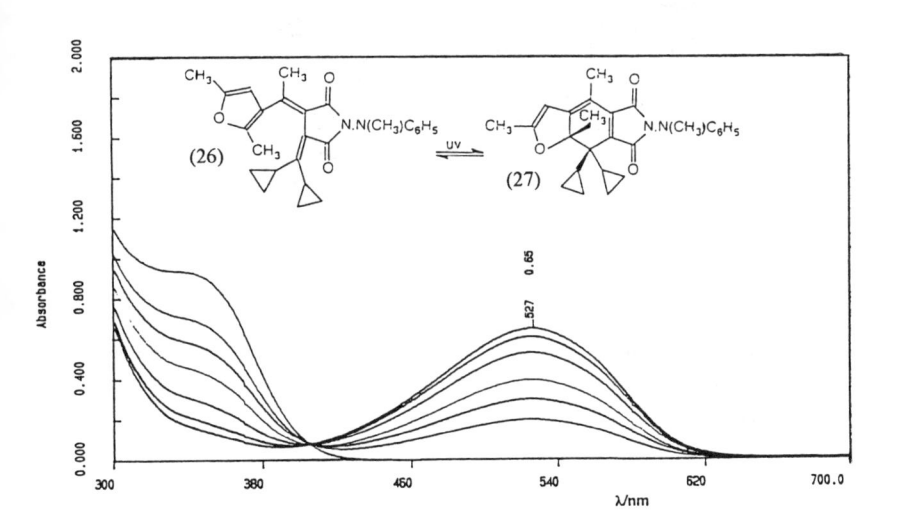

Figure 6 Photochromism of fulgimide (26) in toluene (1x10^{-4}M)

PHOTOCHROMIC DICYANOMETHYLENE DERIVATIVES OF FULGIDES

While substitution of the carbonyl oxygen by nitrogen in a fulgide to form a beta isofulgimide does not have a marked effect on the photochromic properties of the system, replacement of the same oxygen by the dicyanomethylene group does.

The cyclised form (29) of 5-dicyanomethylene-4-dicyclopropylmethylene-3-(2,5-dimethyl- 3-furylethylidene)tetrahydrofuran-2-one (28) shows a bathochromic shift of the maximum of its long wavelength absorption band of 136nm compared to the cyclised form of the corresponding fulgide (14) with a consequential colour change from red to blue (figure 7).

Whether the 2-dicyanomethylene derivative or the 5-dicyanomethylene derivative of the fulgide is obtained depends on which carbonyl group of the fulgide is more susceptible to nucleophilic attack by the dicyanomethane anion.

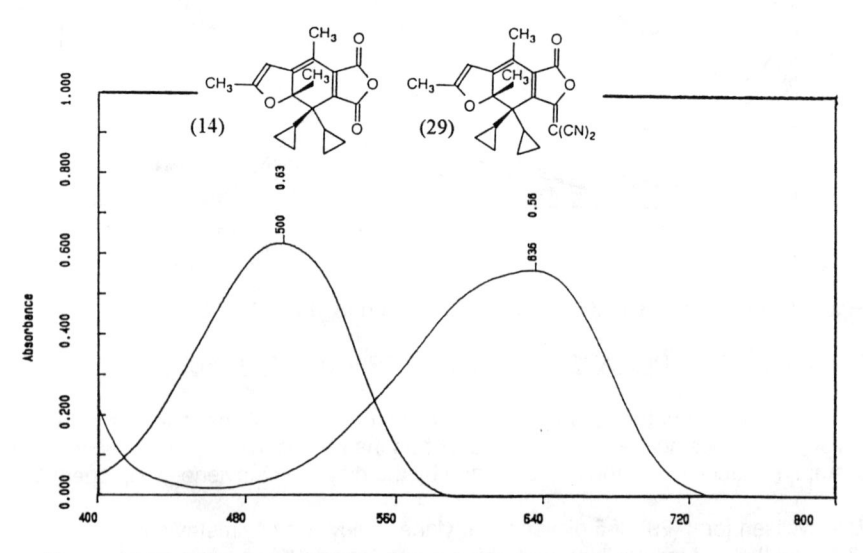

Figure 7 Absorption spectra of cyclised fulgide (14) and its beta dicyanomethylene derivative (29) in toluene (1×10^{-4}M)

When the hetero atom is replaced by sulphur and the 5-methyl substituent of the 3-thienyl group is replaced by a phenyl, the cyclised form (31) of (5-dicyanomethylene-4-isopropylidene-3-(2-methyl-3-phenylthienyl-)-ethylidenetetrahydrofuran-2-one) (30) has a long wavelength absorption band extending into the infrared region (λ_{max} 669nm, ε 16200 dm^3 mol^{-1} cm^{-1}) as shown in figure 8 [cf. cyclised form of fulgide (1), λ_{max} 496nm, ε 8200 dm^{3}mol^{-1} cm^{-1}]. A study of this new infrared active photochromic derivative of a fulgide and related compounds is in progress.

Using procedures similar to those we have described it should be possible to prepare the dicyanomethylene derivative (32, X= C(CN)$_2$) of the photochromic bis-(2-methyl-3-benzothienyl)maleic anhydride (32, X=O) and the related compounds reported by Irie[10].

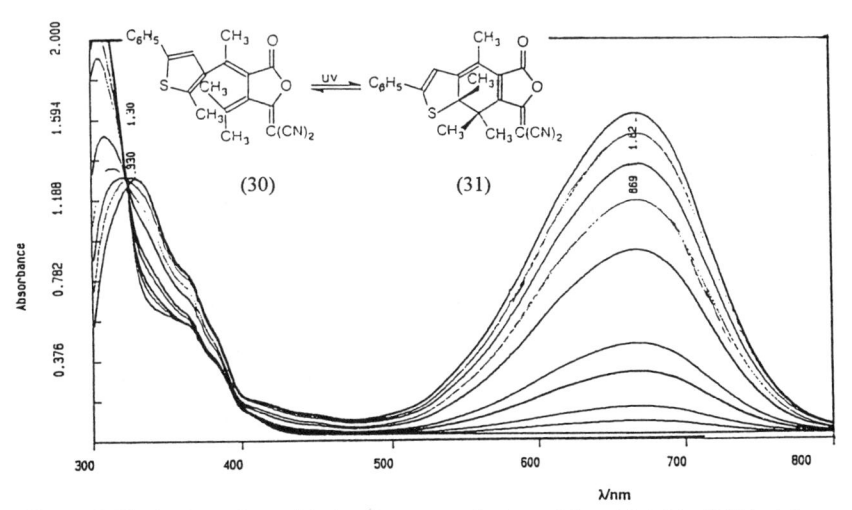

Figure 8 Photochromism of beta dicyanomethylene thienyl fulgide (30) in toluene (1x10^{-4}M).

Optically Active Photochromic Fulgides and Derivatives

In our introduction, it was mentioned that non-destructive read out in optical recording systems may take advantage of optical activity in photochromics. We have prepared optically active fulgimides with a chiral N substituent such as (34) from the reaction of fulgide (1,X= O) with R(+)-1-phenylethylamine. This photodissociates, presumably via a chiral cage mechanism and loses its optical activity on irradiation as shown below:

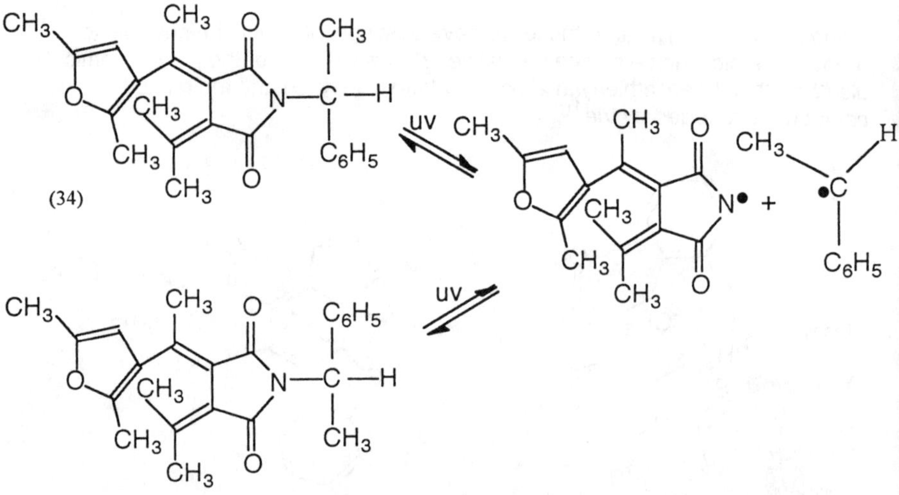

Using a different approach optically active fulgides, in which the optical activity is due in part to their helicity, have been prepared by asymmetric induction by incorporation of the R-3-methylcyclohexanylidene group.

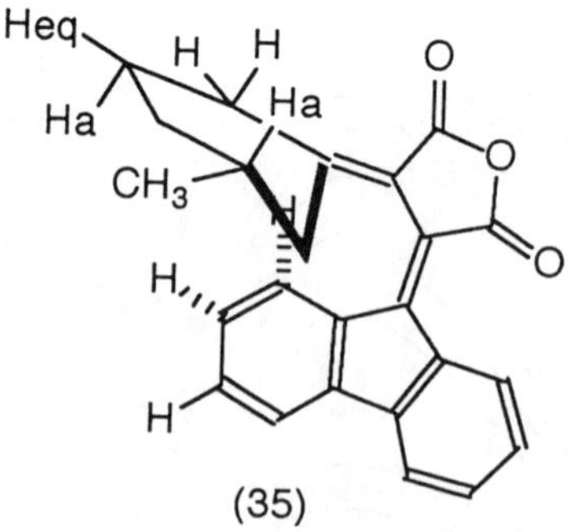

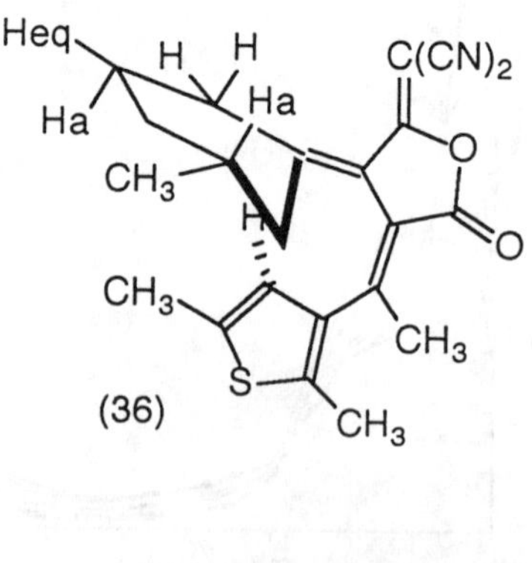

X-ray crystallographic analysis of fulgide (35) shows that the methyl group of the cyclohexanylidene moiety adopts an equatorial position. This conformation results in the fulgide adopting a helical configuration in which the fluorenylidene group is adjacent to the equatorial 2-hydrogen of the 3-methylcyclohexanylidene group and not to the axial 2-hydrogen as can be seen from figure 9.

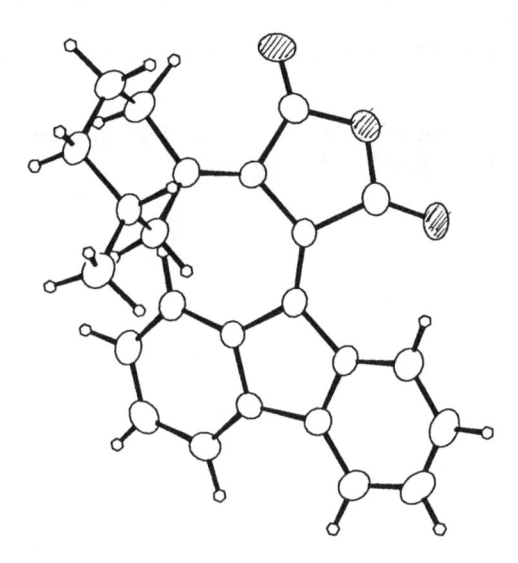

Figure 9 X-ray crystal structure of fulgide (35)

The syntheses of infrared and optically active, thermally stable and fatigue resistant photochromic fulgide derivatives (such as the dicyanomethylene compound (36)), has been accomplished.

Acknowledgements

We thank MTM Research Chemicals Ltd (Lancaster Synthesis) for their generous support of this programme; Faisal University, Saudi Arabia for the research grant to Mr. Shar Al-Shihry, Professor M.B. Hursthouse and Dr. K.M. Abdul Malik for the determinations of structures by X-ray crystallographic analyses.

References

1. H.G.Heller, S.A.Harris and S.N.Oliver, J.Chem.Soc. Perkin 1, 1991, 3258.

2. D.Wood, PhD Thesis , University of Wales , Cardiff 1991.

3. Y.Yokoyama, T.Goto, T.Inoue, M. Yokoyama and Y.Kurita, Chemistry Letters, 1989, 1049.

4. H.G.Heller, A.G.Glaze and J.Whittall, J.Chem.Soc., Perkin 2 ,1992, 591.

5. M.J.Jorgensen and T.Leung, J.Amer. Chem.Soc., 1968, **90**, 3769.

6. Y.Kurita, private communication; B.Helliwell, PhD Thesis, University of Wales, Cardiff 1990.

7. I.Cabrera, D.Achim and H.Ringsdorf, German Patent DE 4007636A1

8. I.Cabrera, A. Dittrich and H.Ringsdorf, Angew. Chem. Engl. 1990, **30**, 76.

9. H.G.Heller, R.J.Hart and K.Salisbury, J.Chem.Soc., Chem Commun., 1968, 1627.

10. M.Irie, S.Nakamura and K.Uchida, Bull. Chem. Soc. Jpn., 1992, **65**, 430-5; M.Irie and A.Mohri, J.Org. Chem., 1988, **53**, 803.

Photochromic Polymers for Optical Data Storage Media

M. Irie

INSTITUTE OF ADVANCED MATERIAL STUDY, KYUSHU UNIVERSITY, KASUGA-KOEN 6-1, KASUGA, FUKUOKA 816, JAPAN

1   INTRODUCTION

Information storage technologies, which provide high capacity, fast access, and low cost memories, are indispensable to a highly developed information-based society. Although magnetic recording presently dominates mass memories, it is being challenged on several fronts by optical recording. Optical recording utilizes focused laser light to effect some optical property change in the recording media which can subsequently be read back by the laser. It has high memory density, because the memory size is determined by the diffraction limit of the laser beam. In addition the optical disks are removable from the drive units. These offer considerable advantages over high density magnetic recording media.

Thus far inorganic media, which use the magneto-optic effect or phase change as the basis of optical recording, have been extensively studied. These media are based on a heat-mode optical recording method. Photon energy of the laser is converted to heat energy on the recording media and used for the recording. In order to utilize the versatile function of light fully, a photon-mode recording method and media should be explored. Photon-mode recording offers advantages with regard to resolution, speed of writing, and multiplex recording capability. It can avoid fatigue due to material movement, and thermal induced damage is reduced. One of the candidates for the photon-mode erasable recording media is organic photochromic materials.

Photochromism is defined as a reversible transformation in a chemical species between two forms having different absorption spectra by photoirradiation. The instant image forming property without processing has led to the consideration of their use in rewritable direct read after write systems. Although much effort has been made in this area in the past, the photochromic materials still await practical application. The limitation is due to the lack of suitable materials which fulfil the requirements for the reversible recording media. Amongst these requirements for the use are:

1) Archival storage capability (thermal stability)
2) Low fatigue (can be cycled many times without of loss of performance)
3) High sensitivity at diode laser wavelengths and rapid response
4) Non-destructive read-out capability.

Perhaps the more important requirements are thermal stability of both isomers, and fatigue resistant characteristics. We recently developed a new type of fatigue resistant and thermally stable photochromic compounds, diarylethenes with heterocyclic rings,[1,2,4-11] where "a" refers to open-ring form and "b" to closed-ring form hereafter.

a   b

For the practical application, however, other properties are also equally required. In this paper, we report on our approach to gain access to a photochromic dye-polymer system which fulfills all of these requirements simultaneously.

2 THERMAL STABILITY

The most important property which photochromic compounds should have for optical data storage media is thermal stability of both isomers. To establish a guiding principle for the thermal stability, we first carried out a theoretical study of 1,3,5-hexatriene to cyclohexadiene type photochromic reactions.[2] The state correlation diagrams were calculated by the semiempirical MNDO method.[3] From the diagrams it was concluded that the thermal stability of both isomers of the diarylethene type photochromic compounds is attained by introducing aryl groups which have low aromatic stabilization energy such as furan or thiophene rings.

The theoretical prediction was confirmed by the synthesis of diarylethenes with various types of aryl groups. Figure 1 shows the thermal stability of photogenerated closed-ring forms of diarylethenes at 80 °C in toluene. The stability depended on the type of aryl groups. When the aryl groups were furan or thiophene rings, which have low aromatic stabilization energy, the closed ring forms were thermally stable. On the other hand, photogenerated closed-ring forms of diarylethenes with phenyl or indole rings were thermally unstable. The photogenerated yellow colour of 9b disappeared with a half lifetime of 1.5 min at 20 °C. The closed-ring form returned quickly to the open-ring form. Compounds 7b and 8b also exhibited thermally reversible reactions.

The different behavior in the thermal stability between diarylethenes with furan or thiophene rings and phenyl or indole rings agrees well with the theoretical

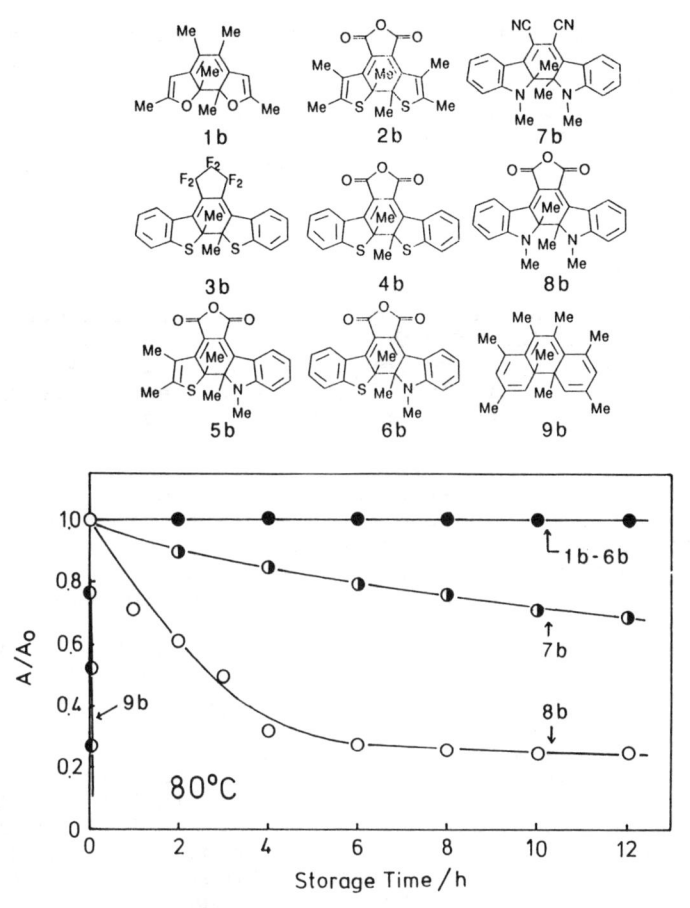

FIGURE 1 Thermal stability described by relative absorbance of photogenerated closed-ring forms of diarylethenes with various types of aryl groups as a function of time.

prediction that the thermal stability depends on the aromatic stabilization energy of the aryl groups.

It is worthwhile noting that the closed-ring forms of 5b and 6b were found to be thermally irreversible but photochemically reversible. The result indicates that the closed-ring forms of non-symmetric diarylethenes are thermally stable when at least one of the heterocyclic rings has low aromatic stabilization energy. In the non-symmetric diarylethene with an indole ring on one end and a thiophene ring on the other end, the thiophene ring plays a role to keep the thermal stability, and the indole ring to shift the absorption band to longer wavelengths (see Section 4).

3 FATIGUE RESISTANT CHARACTER

Photochromic reactions are always attended by rearrangement of chemical bonds. During the bond rearrangement undesirable side reactions take place to some extent. This limits the cycles of photochromic reactions. The difficulty in obtaining fatigue resistant photochromic compounds can be easily understood by a simple calculation. We assumed the following reaction scheme, in which a side reaction to produce B' is involved in the forward process.

$$B' \xleftarrow{\Phi s} A \rightleftarrows B$$

Even under the conditions that the side reaction quantum yield, Φs, is as low as 0.001 and B perfectly converts to A, 63 % of the initial concentration of A will decompose after 1,000 colouration/decolouration cycles. Thus the quantum yield for conversion to byproducts should be less than 0.0001 in order to repeat the cycles more than 10,000 times.

With the aim of clarifying photochemically robust structures, we measured the fatigue resistance property of diarylethenes with various types of aryl groups. Figure 2 shows an example of the fatigue resistant behavior. A benzene solution containing 2,3-bis(2-methylbenzothiophen-3-yl)maleic anhydride 4a(3.7 x 10^{-4} M in a thin cell with pass length of 2 mm) was irradiated alternatively with 436 nm and 546 nm light under deaerated conditions. The absorption intensity of the closed-ring form remained almost constant even after 6 x 10^3 cycles. The intensity decreased to 80 % after 10^4 cycles.

Table 1 summarizes the result of repeatable cycle numbers. The repeatable cycle number indicates when the coloured intensity decreases to 80 % of the first cycle. For the symmetric diarylethenes, the repeatable cycle number was limited to less than 480 times even in the absence of air so far as the compounds had thiophene rings.

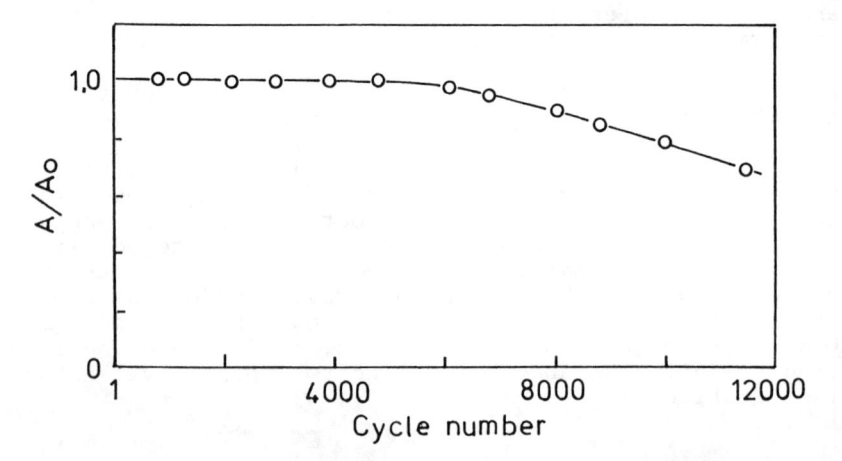

FIGURE 2 Fatigue resistant properties of 4 in deaerated benzene solution following alternative irradiation at 436 nm and 546 nm. The concentration was 3.7 x 10^{-4} M.

Table 1 Fatigue resistant properties

Compd	Repeatable Cycle Number in air	under vacuum
(NC, CN thiophene derivative)	10	—
(maleic anhydride thiophene)	70	480
(maleic anhydride benzothiophene)	3.7 x 10³	1.0 x 10⁴
(maleic anhydride indole-thiophene)	—	8.7 x 10³
(maleic anhydride bis-indole benzothiophene)	—	>1.1 x 10⁴
(perfluorocyclopentene benzothiophene)	>1.3 x 10⁴	—

When the thiophene rings were replaced with benzothiophene rings, the number remarkably increased. Non-symmetric diarylethenes with an indole ring on one end had good fatigue resistance properties. 2-(1,2-Dimethyl-3-indolyl)-3-(2,4,5-trimethyl-3-thienyl)maleic anhydride 5a for example, kept the adequate photochromic property even after 8.7 x 10³ cycles. Diarylperfluorocyclopentene derivatives, such as 1,2-bis(2-methylbenzothiophen-3-yl)-perfluoropentene 3a, showed excellent fatigue resistance properties even in the presence of air.[9]

4 SENSITIVITY AT LONGER WAVELENGTHS

Although recently remarkable progress has been made in diode laser technologies, the wavelength of high power lasers is still longer than 650 nm. From the view point of practical application the photochromic compounds should have the sensitivity in the region of wavelengths 650 - 830 nm. In order to obtain compounds which have sensitivity at longer wavelengths, symmetric diarylethenes with various heterocyclic rings have been synthesized.[4,6,7,10] Among the compounds the closed-ring form of 2,3-bis(1,2-dimethyl-3-indolyl)maleic anhydride 8b had the longest absorption maximum at 620 nm. It was, however, thermally unstable and returned to the open-ring form in the dark, as shown in Fig. 1. In an attempt to get a thermally stable photochromic compounds having the absorption band at longer wavelengths we synthesized unsymmetrically substituted non-symmetric diarylethenes having an indole

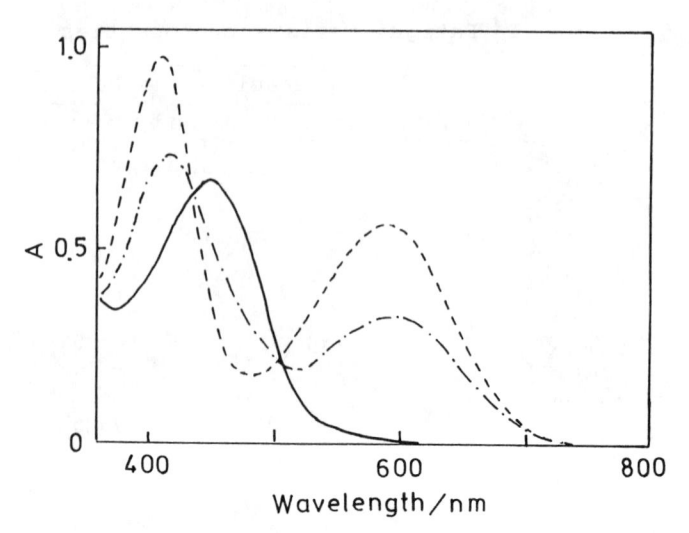

FIGURE 3 Absorption spectra of 5a (—), 5b (---) and the spectrum at the photostationary state under irradiation with 491 nm light (-·-) in benzene.

ring on one end and a thiophene or a benzothiophene ring on the other end.

Figure 3 shows the spectra of 5a, 5b, and the spectrum at the photostationary state under irradiation with 491 nm light in benzene. Upon irradiation with 491 nm light in the presence of air, the solution turned green and a new peak appeared at 595 nm. Isosbestic points were observed at 437 and 504 nm. On exposure to visible light ($\lambda > 560$ nm), the solution again became yellow, and the initial absorption of 5a was restored. The photogenerated closed-ring form was stable, and maintained a constant absorption intensity for more than 12 h at 80 °C. Re-

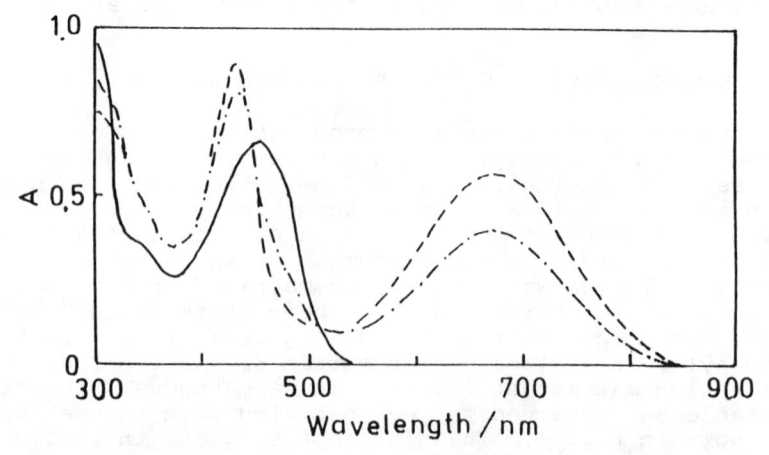

FIGURE 4 Absorption spectra of 12a (—), 12b (---) and the spectrum at the photostationary state under irradiation with 490 nm light (-·-) in hexane.

Table 2 Absorption maxima of non-symmetric diarylethenes in hexane

Structure		λmax	Conversion,%
	5b	578 nm	70
	6b	583 nm	66
	10b	611 nm	71
	11b	626 nm	78
	12b	680 nm	70

placing a thiophene ring of symmetric bis(2,4,5-trimethyl-3-thienyl)maleic anhydride 2a with an indole ring caused a bathochromic shift of the closed-ring form as much as 45 nm. The maximum, however, was shorter than that of 8b.

To shift the maximum further to longer wavelengths, we introduced an electron donating group into the indole ring and an electron withdrawing group into the thiophene ring. The closed-ring forms of 2,3-diarylmaleic anhydride with a 5-methoxyindole ring on one end and a thiophene ring on the other end 10b had the maximum at 611 nm in hexane, while the derivative with an indole ring and a 5-cyanothiophene ring 11b absorbed at 626 nm. The maxima were still shorter than 650 nm.

Finally, we synthesized 2-(5-methoxy-3-indolyl)-3-(2,4-dimethyl-5-cyano-3-thienyl)maleic anhydride 12a. The absorption maximum of the closed-ring form was observed at 680 nm and the edge was extended to 860 nm, as shown in Fig. 4. Table 2 summarizes the absorption maxima of the derivatives.

5 NON-DESTRUCTIVE READ-OUT CAPABILITY

A remaining problem to be solved for the application to optical recording is a non-destructive read-out. Photochromic media do not have read-out stability. Even very weak light can induce the reactions in proportion to the number of photons absorbed by the media. Therefore, after many read-out operations the memory is destroyed. A property that is desired to avoid the destructive read-out problem is a gated photochemical reactivity. The gated reactivity means the property that irradiation with any wavelength causes no reaction, while a photoreaction occurs when another external stimulation, such as heating or light of another wavelength, is present. We tried to construct a reaction system, in which the photochemical reaction is suppressed at room temperature, while it occurs at higher temperatures.

The photochromic reaction of 2-(2-methoxybenzothiophen-3-yl)-3-(1,2-dimethyl-3-indolyl)maleic anhydride 13a depends on solvent polarity. In hexane an effective ring-closure reaction was observed, while in benzene the reaction was strongly suppressed. The conversion from the open-ring to closed-ring form in hexane was over 90 %, while it decreased to less than 5 % in benzene. The solvent dependence is ascribable to the decrease in the ring-closure quantum yield in polarizable solvents. Fluorescence measurement revealed that there existed two kinds of conformations in the excited states, reactive planar and inactive twisted ones, and the inactive twisted conformation became more stable and dominant in polar solvents.[11] This explains the low reactivity in benzene. Similar solvent dependence was observed for the diarylmaleic anhydride with 5-methoxyindole and 5-cyanothiophene 12a. The ring-closure quantum yield in the mixture of hexane and benzene(1 : 1) was less than 10 % of the value in hexane.

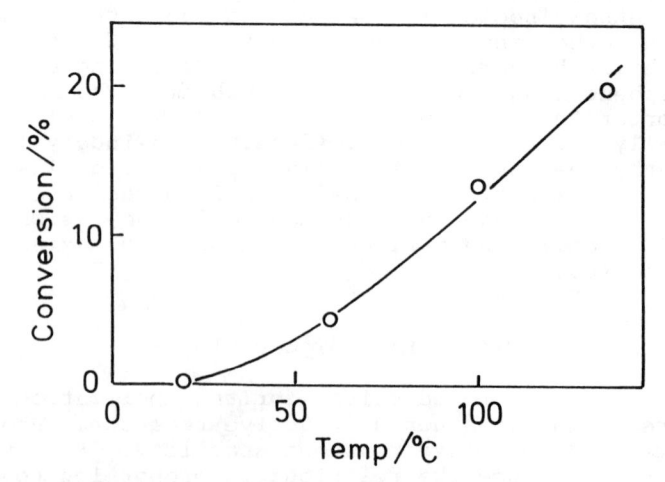

FIGURE 5 Temperature dependence of photochromic reaction of 13a in polystyrene film.

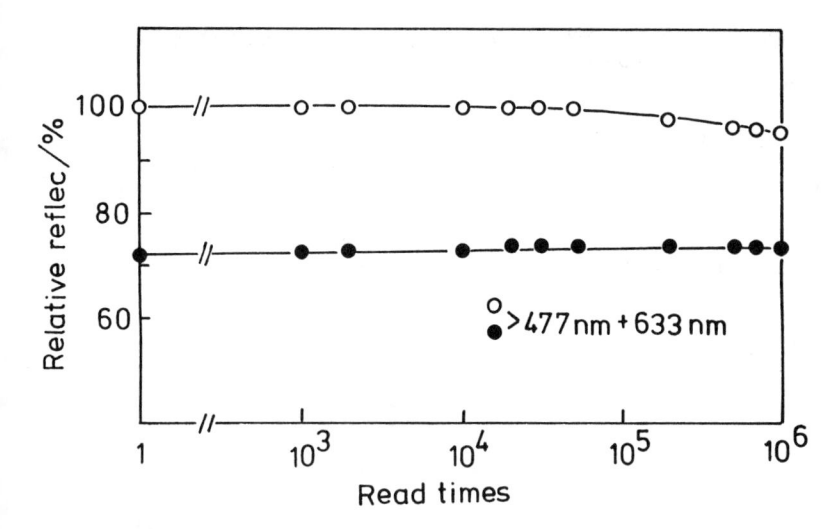

FIGURE 6 Changes of reflection of polystyrene film containing 13a after many times read-out with two lasers, 0.13 mw 477 nm and 0.05 mw 633 nm lasers.

Such a large medium dependence was used for controlling the reactivity by heating. When 13a was dispersed in polystyrene, the compound was photochemically inactive at room temperature. The photocyclization reaction was not observed even by prolonged irradiation with 490 nm light. On the other hand, the compound became photochemically active when the polymer film was heated above 60°C, as shown in Fig. 5. The reactivity change is ascribable to the decrease in dielectric constant of the polymer medium and the increase in the mobility at higher temperature.

The gated photochemical reactivity made it possible to read the memory many times without destruction. Write/read processes were carried out as described below. The initial unrecorded status of the medium is governed by the open-ring form of the photochromic compound which has absorption at 477 nm but none at 633 nm. Writing is initially performed solely with pulsed 477 nm light of sufficiently high intensity to provide the requisite temperature jump of the medium so that photoinduced ring-closure can occur. High contrast between recorded and unrecorded bits can now be achieved by reading simultaneously at the two wavelengths 477 nm and 633 nm, operating in continuous wave mode, but with the former of reduced intensity. Modulation of the read signal results because of the new absorption of the closed-ring form at 633 nm which attenuates light reflection at this wavelength from the reflective coating beneath the photochromic layer. As a result of coupling between the photo-converted form and the read-out wavelength at 633 nm, recorded areas are held at a high temperature relative to unrecorded areas which do not absorb here. The simultaneous irradiation with low intensity 477 light ensures that reversion to open-ring form(ie. erasure) does not occur in recorded data areas by effectively rewriting. However, such treatment cannot effect undesired write events because no thermal bias is provided in unrecorded areas by the 633 nm read beam since absorption at this wavelength is absent. Using this technique reflection modulation was almost constant even after a million cycles. The medium had high read-out stability and write/read/erase cycles could be repeated more than 3×10^3 times without loss of performance.

REFERENCES

1. M. Irie and M. Mohri, J. Org. Chem., 1988, 53, 803
2. S. Nakamura and M. Irie, J. Org. Chem., 1988, 53, 6136
3. M.J.S. Dewar and W. Thiel, J. Am. Chem. Soc., 1977, 99, 4899
4. Y. Nakayama, K. Hayashi and M. Irie, J. Org. Chem., 1990, 55, 2592
5. M. Irie, Jpn J. Appl. Phys., 1989, 28-3, 215
6. K. Uchida, Y. Nakayama, M. Irie, Bull. Chem. Soc. Jpn, 1990, 63, 1311
7. Y. Nakayama, K. Hayashi and M. Irie, Bull. Chem. Soc. Jpn, 1991, 64, 789
8. M. Irie, J. Synth. Org. Chem. Jpn, 1991, 49, 373
9. M. Hanazawa, R. Sumiya, Y. Horikawa, M. Irie, J. Chem. Soc. Chem. Commun., 1992, 206
10. K. Uchida and M. Irie, Bull. Chem. Soc. Jpn, 1992, 65, 430
11. M. Irie and K. Sayo, J. Phys. Chem., 1992, 96, 7671

Photochromic Organic Compounds in Polymer Matrices

John C. Crano,[*] Cletus N. Welch, Barry Van Gemert, David Knowles, and Bruce Anderson

PPG INDUSTRIES, INC., 440 COLLEGE PARK DRIVE, MONROEVILLE, PA 15146, USA

1. INTRODUCTION

Although the proposed applications for photochromic systems are numerous, only one has received broad use, that involving the use of photochromic inorganic glass in prescription eyewear. An estimated 18 million pairs of photochromic prescription lenses were dispensed internationally in 1989. Almost all of these lenses were prepared from mineral glass containing silver/copper halide.

Another significant trend has been the growth of lightweight plastic lenses for eyewear. In the United States, about 75% of prescription lenses are now made of plastic. With this increasing market penetration, the demand for an acceptable plastic photochromic lens has also increased.

Several years ago, PPG Industries set out on a long range research and development program with the goal of discovering plastic photochromic systems which could be used to prepare prescription eyewear. Considerable progress has been made in this continuing quest. With this paper, we would like to review with you some of the results we have achieved in the science and technology of organic photochromic systems. But before doing so, a brief examination of some related literature pertaining to the properties of organic photochromic compounds, especially spiro-oxazines, would be beneficial.

2. BACKGROUND

Since the discovery of the photochromism of indolino spirobenzopyrans by Fischer and Hirshberg in 1952,[1] this family and other related spiropyrans have been studied extensively. Fairly complete reviews of the information generated on the chemistry of spiropyrans have been written, first by Bertelson[2] and later by Guglielmetti.[3]

In general, properly substituted spiropyrans can exhibit excellent photochromic responses. That is, upon exposure t UV radiation, rapid and deep coloration of a solid or liquid solution of the spiropyran occurs. The accepted mechanism for the coloration involves the formation of the photomerocyanine as described in equation 1 for the common spiro pyran, I (6-nitro-1′,3′,3′-trimethylspiro[2H-1-benzopyran-2,2′-indoline]).

$$\text{I} \rightleftarrows \text{(photomerocyanine)} \quad (1)$$

The extent of coloration or optical density at equilibrium is dependent on a number of factors - the intensity of the light source, the quantum efficiency for the forward reaction, the extinction coefficient of the merocyanine, and the rate of the reverse reaction. The reclosure of the pyran ring, the reverse reaction indicated in equation 1, can be driven either thermally or photochemically.

The rate of the thermal reversion of the merocyanine to the spiropyran is controlled by a number of factors - substitution pattern, temperature, etc. One of the most important factors is the matrix in which the compound is dissolved. For example, in a recent paper, Keum et al.[4] described the effect of solvent polarity on the rate of thermal reversion of nitro substituted photomerocyanines. The rate of reversion decreased dramatically with increasing solvent polarity. The rate constant for decoloration of the 6-nitro-8-methoxy spiropyran, II, was 44.2×10^{-3} sec^{-1} in toluene versus 3.45×10^{-3} sec^{-1} in dimethyl sulfoxide.

II

Essentially all of the very responsive indolino spiropyrans are substituted with a nitro group on the benzopyran portion of the molecule. Unfortunately, the nitrospiropyrans have another property which detracts from their utility. In general, the spiropyrans and in particular the nitro substituted compounds have very poor resistance to photochemically induced degradation.

For any utilization of organic photochromic systems in an ophthalmic lens, this problem of photodegradation, or as we call it, fatigue, had to be solved. It was somewhat alleviated with the discovery of indolino spironaphth-oxazines. This family is inherently more resistant to fatigue than other organic photochromic compounds. The extent of this difference is dependent on the method of measurement. As measured by the quantum yield of photodegradation, spiro-naphthoxazines were shown to be two to three orders of magnitude more resistant to fatigue than the nitro-spirobenzopyrans.[5] The stability of the naphthoxazines can be enhanced even further. Their photostability can be improved by protecting them from oxygen or by the addition of chemical stabilizers, specifically, nickel complexes or hindered amine light stabilizers.[6,7]

Now we would like to discuss some of the contributions that we have made at PPG in the area of spiro-oxazine chemistry with special emphasis on their properties in polymeric matrices. This work has been going on for several years and thus qualifies as a long-term research project. Nevertheless, the work has been focussed on the target of a viable plastic photochromic lens system. This will become obvious as we go over some of the procedures used to prepare and test photochromic samples.

3. EXPERIMENTAL

Synthesis of Spiro-oxazines

The syntheses of the spiro-oxazines used in this study will not be covered in detail. The general reactions that were used to prepare the subject compounds is illustrated with the example given in equation 2. In almost all cases, the experimental photochromic compounds were purified by liquid chromatography.

Measurement of Photochromic Properties

One piece of equipment has been used extensively for the measurement of the photochromic response of various samples. The apparatus is based on the use of an Ealing

Triangular optical bench equipped with a 300 watt xenon arc lamp (Oriel 6258) (see Figure 1). Light from the lamp was passed through a copper sulfate solution to remove infrared, a neutral density glass filter to regulate irradiance, a shutter assembly and the sample. Sample optical density was determined using a tungsten lamp, a photopic filter and silicon detector mounted on a separate optical bench positioned at an angle to the first bench.

The photochromic response of the sample was defined as the change in optical density (ΔOD) upon exposure to the light from the xenon lamp as measured using the photopic filter. The change in optical density was determined as a function of time of exposure and both an initial rate of change, calculated from a 30 second reading, and the equilibrium or steady state response were determined for each sample. After achieving the steady state condition, exposure to the activating radiation was discontinued and the rate of decoloration was measured. Usually, the rate of fading was recorded as a half-life, the time to return to ΔOD of one-half the equilibrium value. The use of the half-life was convenient for our comparison of samples but does not imply unimolecular kinetics.

The final photochromic property to be measured was photostability or fatigue resistance. In order to accomplish this, the equilibrium response of each sample was determined and the sample was then continuously exposed to UV radiation within an Accelerated Weathering Tester (The Q-Panel Company) equipped with UVA-351 lamps. After several hours of exposure, the equilibrium response of the sample was measured again. By this procedure, the rate of fatigue could be calculated by determining the reduction of the equilibrium response as a function of time of UV exposure. The fatigue resistance of the sample was usually given as a half-life - the hours of UV exposure to decrease the equilibrium response to one-half that of the original sample.

Preparation of Samples

There are several potential methods for the preparation of photochromic samples for testing. An important criterion for the methods chosen in our work was that they be realistic simulations of practical methods for the preparation of ophthalmic lenses. For several reasons we chose to do most of our studies with samples prepared through an imbibition procedure in which one surface of a plastic piece was penetrated, or imbibed, with the photochromic compound to be tested.

PPG supplies a monomer for the casting of ophthalmic lenses, CR-39^R monomer. CR-39^R monomer is essentially the bis(allyl carbonate) of diethylene glycol. Polymerization of this monomer requires a large amount of peroxidic initiator, such as 2 to 4% of a peroxydicarbonate. In

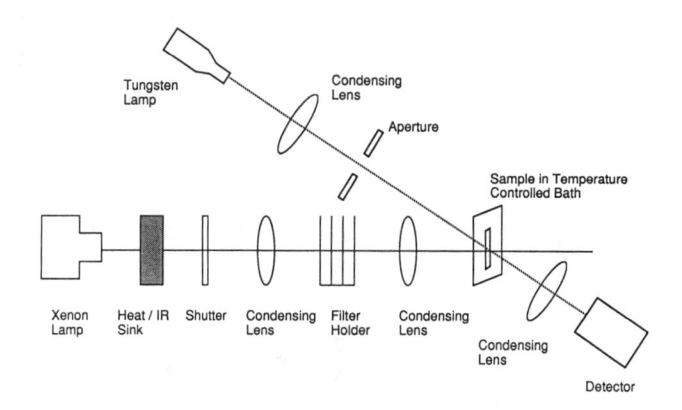

Figure 1

Schematic Diagram of Optical Bench

ttempting to cast plastic pieces from CR-39^R monomer in
he presence of spiropyrans or spiro-oxazines, the
hotochromic compounds were destroyed by the high
oncentration of the free radical initiator.

Photochromic samples of the resin from this monomer
ould be prepared by an imbibition technique.[8] The process
an be illustrated with an example. If a piece of standard
aboratory filter paper is saturated with a solution of a
ompound and then allowed to air dry, the result is a paper
hroughout which the compound is evenly dispersed. By
lacing this paper on the surface of the plastic piece and
eating the assembly, the photochromic compound transfers,
r imbibes into the surface. With variation of the
emperature and time of the imbibition, control of the
mount of photochromic transferred as measured by the
ltraviolet absorption of the sample could be achieved.
are was taken to compare the various photochromics within
class at comparable effective concentrations.

4. RESULTS AND DISCUSSION

For reference purposes, a few indolino spiro-
aphthoxazines were prepared and tested using our
echniques. As stated before, the photochromic compounds
ere imbibed into the surfaces of flat square plates
repared from CR-39^R monomer. The photochromic properties
f some of these naphthoxazines are presented in Table 1.
he characteristic spectrum of the open form of one of the
aphthoxazines, III, is given in Figure 2.

In Table 2 are compared the photochromic properties of
wo isomeric naphthoxazines, one derived from 1-nitroso-2-

TABLE 1

PHOTOCHROMIC PROPERTIES OF SUBSTITUTED INDOLINO SPIRONAPHTHOXAZINES

COMPOUND	R_2	SUBSTITUENTS R_3	R_4	R_5	SENSITIVITY (ΔOD/Min)	EQUILIBRIUM RESPONSE (ΔOD)	FADE T ½ (Sec)	FATIGUE T ½ (Hours)	λ max ACTIVATED (nm)
III	CH_3	CH_3	CH_3	--	0.33	0.10	24	345	607
IV	CH_3	CH_3	5,6(or 4-) $(CH_3)_2$	--	0.44	0.22	58	410	615
V	-$(CH_2)_5$-		--	--	0.47	0.26	87	340	613
VI	CH_3	CH_3	5-Cl	--	0.26	0.08	32	200	606
VII	CH_3	CH_3	5,6(or 4-) $(CH_3)_2$	9'-OCH_3	0.21	0.19	124	205	606
VIII	C_2H_5	C_2H_5	--	9'-OCH_3	0.22	0.21	147	90	606
IX	CH_3	CH_3	--	6'-N $(CH_2)_5$	0.64	0.27	58	50	583
X	CH_3	CH_3	--	6'-N $(CH_2)_5$ 9'-OCH_3	0.45	0.37	117	30	578

TABLE 2

PHOTOCHROMIC PROPERTIES OF INDOLINONAPHTHOXAZINES, IV AND XI

COMPOUND	SENSITIVITY (ΔOD/Min)	EQUILIBRIUM RESPONSE (ΔOD)	FADE T½ (Sec)	FATIGUE T½ (Hours)	λ max ACTIVATED (nm)
IV	0.44	0.22	58	410	615
XI	0.36	0.29	103	350	601, 637

naphthol, IV, the other from 2-nitroso-1-naphthol, XI. Their properties are surprisingly similar except for the visible spectra of the corresponding photomerocyanines. Compound IV has the spectrum considered typical for an indolino spironaphthoxazine with a principle absorption band in the region of 600 to 620 nm plus a shoulder at around 570 nm. The 2-nitroso-1-naphthol derivative, XI, has two absorption bands of nearly equal intensity at 601 and 637 nm.

A few years ago, Kwak and Hurditch[9] patented the family of indolino spiro-oxazines derived from 5-nitro-6-hydroxyquinoline. This family, the spiropyridobenzoxazines is similar to the spironaphthoxazines in the visible absorption bands of the open forms but, in general, possess greater sensitivities and equilibrium responses (see Table 3). For example, compounds IV and VII, both naphthoxazines and compound XII, a pyridobenzoxazine, have the same substitution pattern on the indolino portion. Nevertheless XII has a sensitivity of 0.56 (ΔOD/min) and an equilibrium response of 0.42 (ΔOD) versus sensitivities of 0.44 and 0.21 and equilibrium responses of 0.22 and 0.19 for IV and VII, respectively. Again, the photochromic properties were measured under the same conditions and at comparable concentrations within the surface of polymer samples prepared from CR-39[R] monomer.

In Figure 3 is given the ultraviolet absorption spectrum of the pyridobenzoxazine, XII. This spectrum, taken in a diglyme solution, is typical of the spectra of compounds within this family of photochromic compounds. Also, the spectrum in the matrix used for most of these studies, the polymer from CR-39[R] monomer is very similar to that in diglyme solution.

Within the family of spiropyridobenzoxazines, a few were prepared with a methyl group replacing the hydrogen normally in the 2'-position, for example, compound XIX. The introduction of this methyl group invariably had the effect of completely eliminating the photochromic character of the compound. It has been suggested that the added methyl group inhibits the formation of a stable planar open form.[10]

XIX

The preparation of the indolino spiro-oxazines from o-nitrosophenols was recently disclosed.[11] Table 4 lists the photochromic properties of various spiro-oxazines

TABLE 3
PHOTOCHROMIC PROPERTIES OF SUBSTITUTED INDOLINO SPIROPYRIDOBENZOXAZINES

COMPOUND	R_1	R_2	R_3	R_4	SENSITIVITY (ΔOD/Min)	EQUILIBRIUM RESPONSE (ΔOD)	FADE T ½ (Sec)	FATIGUE T ½ (Hours)	λ max ACTIVATED (nm)
XII	CH_3	CH_3	CH_3	5,6(or 4-)(CH_3)$_2$	0.56	0.42	122	280	609
XIII	CH_3	CH_3	C_2H_5	5,6(or 4-)(CH_3)$_2$	0.64	0.53	181	---	610
XIV	n-C_3H_7	CH_3	CH_3	H	0.44	0.21	60	425	605
XV	CH_3	CH_3	CH_3	5-OCH_3	0.37	0.32	127	---	617
XVI	C_2H_5	CH_3	CH_3	5-OCH_3	0.50	0.37	106	---	618
XVII	n-C_3H_7	CH_3	CH_3	5-OCH_3	0.50	0.38	117	160	619
XVIII	n-C_4H_9	CH_3	CH_3	5-OCH_3	0.51	0.38	116	---	615

TABLE 4
PHOTOCHROMIC PROPERTIES OF INDOLINO SPIROBENZOXAZINES

COMPOUND	SUBSTITUENTS R₄	R₅	SENSITIVITY (ΔOD/Min)	EQUILIBRIUM RESPONSE (ΔOD)	FADE T½ (Sec)	FATIGUE T½ (Hours)	λ max ACTIVATED (nm)[a]
XX	H	5,7-(OCH₃)₂	0.27	0.20	66	100	463, 558
XXI	H	6,7-(OCH₃)₂	0.54	0.27	62	160	434, 584
XXII	H	5,7-(OCH₃)₂, 8-CH₃	0.25	0.19	75	30	489
XXIII	5',6'(or 4'-)(CH₃)₂	5,7-(OCH₃)₂	0.26	0.31	170	185	462, 566
XXIV	5',6'(or 4'-)(CH₃)₂	6,8-(t-C₄H₉)₂	0.02	0.02	103	---	---
XXV	5',6'(or 4'-)(CH₃)₂	7-N(C₂H₅)₂	1.01	0.39	69	---	565
XXVI	5'-OCH₃	5,7-(OCH₃)₂	0.29	0.33	225	110	468, 570
XXVII	5'-NO₂	5,7-(OCH₃)₂	1.37	-0.50	140	<5	479, 545
XXVIII	4'-(or 6'-)NO₂	5,7-(OCH₃)₂	1.34	0.45	79	30	463, 540
XXIX	5'-CN	5,7-(OCH₃)₂	0.15	0.10	68	70	468, 540
XXX	5'-F	5,7-(OCH₃)₂	0.22	0.20	87	---	465, 558

(a) MORE INTENSE ABSORPTION BAND UNDERLINED

belonging to this family. Because of their photochromic properties and relatively good photostability, most of our studies were concentrated on the methoxy substituted benzoxazines. One interesting property of each of these compounds is the visible spectrum of the corresponding photomerocyanine. In most cases, two absorption bands are observed, the positions and relative intensities being controlled by the substitution pattern. Figures 4, 5, and 6 are the visible spectra of the open forms of three methoxy substituted indolino spiro-benzoxazines.

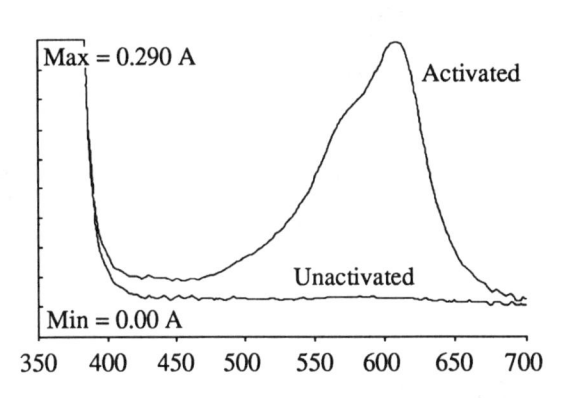

Figure 2

Visible Absorption Spectrum of Open Form of
1,3,3-Trimethylspiro[indoline-2,3'-[3H]naphth-[2,1-b][1,4]oxazine] (III)

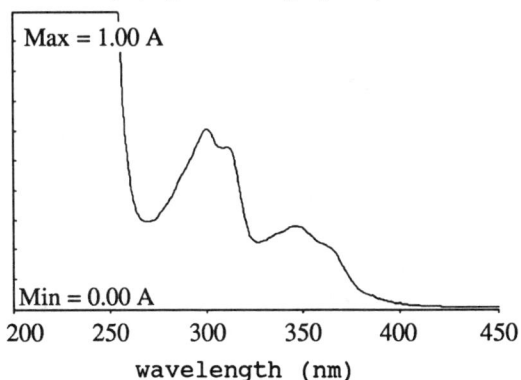

Figure 3

UV Spectrum of 1,3,3,5,6(or 4-)Pentamethyl spiro[indoline-2,3'-[3H]pyrido[3,2-f]-[1,4]benzoxazine] (XII) (Diglyme Solution)

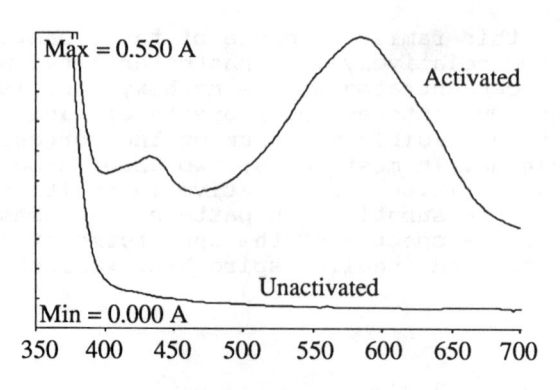

Figure 4

Visible Absorption Spectrum of Open Form of
6,7-Dimethoxy-1',3',3'-trimethylspiro[2H-1,4-benzoxazine-2,2'-indoline] (XXI)

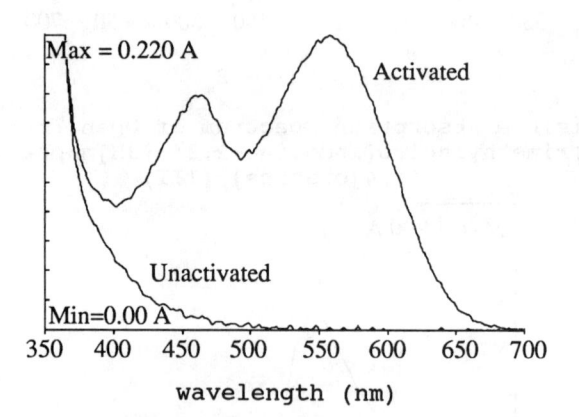

Figure 5

Visible Absorption Spectrum of Open Form of
5,7-Dimethoxy-1',3',3'-trimethylspiro[2H-1,4-benzoxazine-2,2'-indoline] (XX)

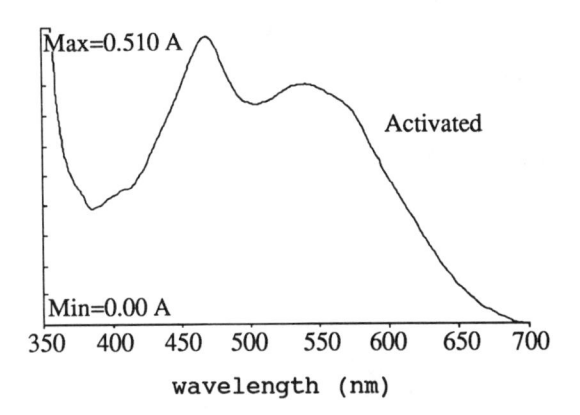

Figure 6

Visible Absorption Spectrum of Open Form of
5,7-Dimethoxy-1',3',3'-trimethyl-5'-cyanospiro[2H-1,4-benzoxazine-2,2'-indoline] (XXIX)

The ultraviolet spectra of most of the 5,7-dimethoxy spirobenzoxazines are similar. There is, in each case, an almost symmetrical absorption band with a maximum at around 300nm.

The nitro-substituted compounds, as might be expected, are very different from the others. The 5'-and 6'-nitro indolino compounds are much more active photochromic compounds with regard to sensitivity or equilibrium response. The activity is accompanied, however, with very poor photostability, especially with the 5'-nitro compound which could be observed to fatigue as the initial photochromic properties were being determined.

In order to examine the effects of substituents on the structures of indolino spirobenzoxazines, the structures of three compounds, XX, XXVI, and XXX, were determined by X-ray diffraction analysis. The structure of the 5'-methoxy substituted compound, XXVI, is shown in Figure 7. As can be seen from the data in the preceding table, the photochromic properties of XX (unsubstituted indolino group) and XXX (5'-fluoro) are similar while those of the 5'-methoxy compound are quite different. It is interesting to note the differences in the lengths of the spiro carbon-nitrogen and spiro carbon-oxygen bonds as measured for the three compounds (Table 5). As can be seen from the data, the length of the spiro carbon-oxygen bond in the methoxy substituted compound is slightly longer than the lengths of the same bond in the other two compounds. The sensitivity and equilibrium response of this compound are also enhanced relative to those of the others.

Table 5

Bond Lengths of Substituted Spirobenzoxazines

Compound	Substituent	Bond Lengths (Å) Spiro C-N	Spiro C-O
XX	5'-H	1.458	1.455
XXVI	5'-CH$_3$O	1.448	1.474
XXX	5'-F	1.448	1.457

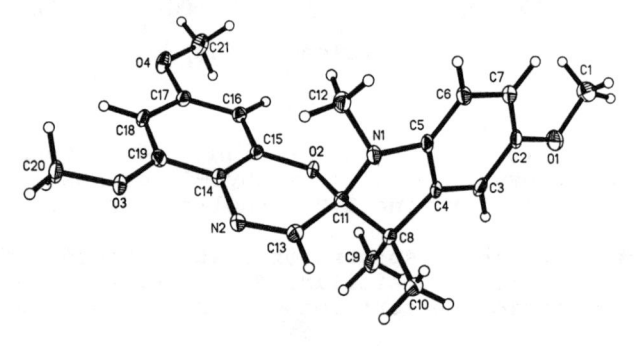

Figure 7

X-ray Structure of Crystalline
5',6,7-Trimethoxy-1',3',3'-trimethylspiro[2H-1,4-benzoxazine-2,2'-indoline] (XXVI)

As might be expected, the photochromic properties of each compound are subject to modification through a change in the matrix. A terpolymer of CR-39[R] monomer (68.5 wt%), Uvithane[R] 893, an oligomeric urethane acrylate (30 wt%), and allyl methacrylate (1.5 wt%) was prepared and used as the matrix for the spirobenzoxazine, XXIII, in the same manner in which the polymer from CR-39[R] monomer was used. At comparable amounts of photochromic compound, the equilibrium response was enhanced in the terpolymer matrix relative to the homopolymer (see Table 6).[12]

The equilibrium responses listed in Table 6 were determined with a different apparatus than that used for the data appearing in the preceding tables. In this case, each sample was irradiated with light from a xenon arc solar simulator (Oriel Model 81172) until an equilibrium luminous transmission was achieved as measured with a color spectrophotometer (Spectrogard II). The initial and final luminous transmissions were used to calculate the changes in optical density.

Table 6

Photochromic Response of Spirobenzoxazine, XXIII, in Homopolymer of CR-39^R Monomer and Terpolymer with Uvithane^R 893 and Allyl Methacrylate

Matrix	Equilibrium Response (ΔOD/min)		
	10°C	22°C	35°C
Homopolymer	0.31	0.16	0.059
Terpolymer	0.43	0.22	0.081

REFERENCES

1. E. Fischer and Y. Hirshberg, *J. Chem. Soc.*, 1952, 4522.

2. R. Bertelson, 'Techniques of Chemistry-Photochromism', Wiley, London, 1971, Vol. 3, Chapter 3, p. 45.

3. R. Guglielmetti, 'Studies in Organic Chemistry, 40 - Photochromism, Molecules and Systems', Elsevier, Amsterdam, 1990, Chapter 8, p. 314.

4. S.-R. Keum, M.-S. Hur, P. M. Kazmaier, and E. Buncel, *Can. J. Chem.*, 1991, 69, 1940.

5. N. Y. C. Chu, 'Proceedings of the 10th IUPAC Symposium on Photochemistry, Interlaken', 1984.

6. N. Y. C. Chu, *U. S. Patent* 4,440,672, 1984, (to American Optical).

7. N. Y. C. Chu, 'Optical Materials Technology for Energy Efficiency and Solar Energy Conversion VII', SPIE, Vol. 1016, p. 152.

8. L. Le Naour-Sene, *U. S. Patent* 4,286,957, 1981, (to Essilor).

9. W. S. Kwak and R. Hurditch, *U. S. Patent* 4,637,698, 1987, (to PPG Industries, Inc.).

10. W. Clegg, N. Norman, T. Flood, L. Sallans, W. S. Kwak, P. L. Kwiatkowski, and J. G. Lasch, *Acta Cryst.*, 1991, C47, 817.

11. W. S. Kwak and C. W. Chen, *U. S. Patent* 4,816,584, 1989, (to PPG Industries, Inc.).

12. D. S. McBain and J. C. Crano, *U. S. Patent* 4,994,208, 1991, (to PPG industries, Inc.).

Experimental and Theoretical Studies on the Electrical Fatigue of Organic Photoconductors

J. Pacansky,[*] R. J. Waltman, R. J. Berry, and J. Carter

IBM ALMADEN RESEARCH CENTER, K35/802, 650 HARRY ROAD, SAN JOSE, CA 95120–6099, USA

1 INTRODUCTION

In recent years, electrophotography has manifested itself as the technology of choice in the area of non-impact printing, spearheaded first by the development of the laser printer in 1976, followed by the application of the diode laser printhead thereafter. The success in each evolution of non-impact printers based upon electrophotography could only be made possible by the concomitant development of the key element in these devices, the photoconductor. The photoconductor is, of course, the photosensitive element that bears the latent electrostatic image that is produced by light, developed and transferred to paper by the toner, and this sequence is repeated many times, and is referred to as the electrophotographic cycle.[1] While the first commercially successful photoconductors used amorphous chalcogenide alloys, most preparations today utilize organic materials. The overwhelming reason is one of cost performance. Organic photoconductors are cheaply fabricated and offer great flexibility in spectral tuning (i.e. wavelength response), making them photosensitive to, and therefore compatible with, a wide variety of printheads, e.g., diode or gas lasers (GaAlAs, $\simeq$830 nm; HeNe, $\simeq$ 630 nm), lamps (visible), light emitting diodes and so on. Since they are fabricated primarily from plastics, they are also mechanically flexible, consequently, they are virtually exclusively used in small belt printers. The major drawback with organic photoconductors is their ultimate lifetime, which can be considerably less than the chalcogenide alloy-based or amorphous silicon-based photoconductors. For example, photoconductors derived from Se alloys may endure as much as $\simeq$300,000 electrophotographic cycles; amorphous silicon $\simeq$1,000,000 cycles; organic photoconductors at best yield $\simeq$100,000 cycles.[2] At the high end of the printer marketplace, a 100,000 cycle lifetime for photoconductors requires their frequent replacement; thus the cost of supplies becomes an important consideration for many businesses.

Here at the IBM Almaden Research Center, we have been interested in elucidating the origins or underlying causes that place a limit on the number of cycles a particular photoconductor formulation may endure. While we note that photoconductors are subjected to a variety of environmental stresses during their lifetime, for example, exposure to corona discharges, heat (fuser), discharge light, and room light (working environment), we have particularly been interested in the effect of light on photoconductors and how this might effect and degrade initial electrical properties, resulting in photoconductor "fatigue." During the course of our studies, we have observed that many organic molecules responsible for carrier generation and transport are photosensitive and can initiate and/or undergo photochemistry when exposed to visible light.[3-6] This occurs during the electrophotographic cycle, or during the manufacture, storage and replacement of photoconductors in the standard work environment. Photochemistry results in chemical changes in the photoconductor which in turn affects their electrical properties. It has become clear that light on photoconductors is a primary cause of electrical fatigue. Nevertheless, their cost performance compensates for many of these deficiencies, rendering organic photoconductors still as the materials of choice in printers and copiers. It is therefore important to improve not only the performance but also the ultimate lifetime of the organic photoconductors, particularly in light of recent industrial targets mandating $\simeq$ 500,000 cycles.[7] Achieving this goal is formidable indeed because the manufacture of a printer requires multi-disciplinary skills, including engineering required to design the printers tailored to the particular photoconductors being used, and chemistry and physics to optimize the performance of the photoconductors. To realize a significant improvement in the organic photoconductor certainly provides a practical problem with considerable scientific challenges. In order to meet these challenges, an understanding of the causes of photoconductor fatigue is first required.

It is the intent of this paper to provide an understanding of the underlying causes for photoconductor fatigue that lead to their shortened lifetimes. The discussions will be limited to the effect of light on photoconductors and we will describe the symptons, causes and resolutions applicable, depending upon the origin of photoconductor fatigue. This entails a detailed discussion of the photochemistry in both the hole transport and carrier generation layers and how the chemical changes induced by light effect electrical properties. In the next several sections, brief discussions on the electrophotographic process and composition of organic photoconductors are presented first to acquaint the interested reader.

The Electrophotographic Process

All modern photocopiers and many printers use an electrophotographic process to prepare hard copies. As stated above, the electrophotographic process involves the placement of charges, negative or positive ions, on the surface of the photoconductor. The charges are selectively discharged by light to produce a latent electrostatic image that is developed by the toner and transferred to paper. This sequence is depicted in Fig. 1 for a typical photoconductor that is wrapped about a cylindrical drum. In step (a), the photoconductor is charged in the dark with negative ions by a corona to about 900 Volts. This is referred to as the dark voltage. The time required for the photoconductor to move from the charging station (a) to the discharge station (b), is $\simeq$ several hundred milliseconds. During this time, the electrical conductivity of the photoconductor must be small enough so that the surface potential changes insignificantly. Any decay in the number of surface charges is referred to as the dark decay. At station (b) areas of the photoconductor are selectively discharged by photo-induced current from a light source, the remaining charges from the non-exposed areas providing a latent electrostatic image. In the areas where discharge has occurred, the surface potential should be reduced to ambient voltages ($\simeq$ 50 Volts) and this process is called light decay. At the center of Fig. 1, the relevant changes in electrical properties of the photoconductor are plotted as a function of time, where the photoconductor is moved from one station to the next. This is a photo-induced discharge curve and is characteristic of ideal behaviour. It follows that any significant deviation in the dark or light decay portions of the curve would represent non-ideal behaviour characteristic of a photoconductor that is "electrically fatigued." In step (c) the latent electrostatic image is developed with toner and in step (d), the developed image is transferred to paper. This cycle is repeated $\simeq$100,000 times over the lifetime of a typical (organic) photoconductor.

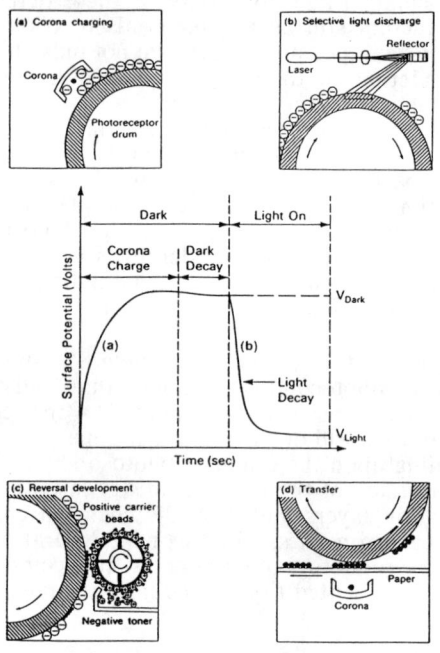

Figure 1. Photoinduced discharge curve related to various points in the electrophotographic process.

It is apparent in Fig. 1 that the photoconductor is subjected to light and corona discharges. Additional environmental stresses not obvious from Fig. 1 are heat, from the fuser roll, and light from fluorescent and/or tungsten light in the work environment, including any exposures during the fabrication of the photoconductor. Although the accumulative incident dose received from room lighting and light used to discharge the photoconductor is small, of the order of several J/cm^2, nevertheless the photocurrents generated during the photodischarge are also very small and subject to significant changes if the incident dose is translated into photochemical changes. As will be discussed in detail later, these very small doses of light cause enough photochemistry in the photoconductor components to induce significant changes in the electrical properties of the photoconductors.

Composition of an Organic Photoconductor

A brief summary of the chemical compostion and structure of an organic layered photoconductor is presented here. As shown in Fig. 2, a typical organic photoconductor is composed of several key layers. An aluminum coated ($\simeq$300 Å) polyethylene terephthalate serves as a substrate and electrical ground. A binder layer, a polyester, is coated atop the aluminized polymer substrate and serves to adhere the next layer, the carrier or charge generation layer (CGL). The CGL is composed of an organic dye like chlorodiane blue which absorbs visible light in the presence of a strong electric field to generate carriers. Atop the charge generation layer is the charge transport layer (CTL), composed of a charge transport molecule such as the hydrazone DEH, dispersed typically to 40% by weight in a polymeric matrix. While Fig. 2 is representative of typical commercial formulations for organic layered photoconductors, a variety of materials may be used for charge generation and transport.

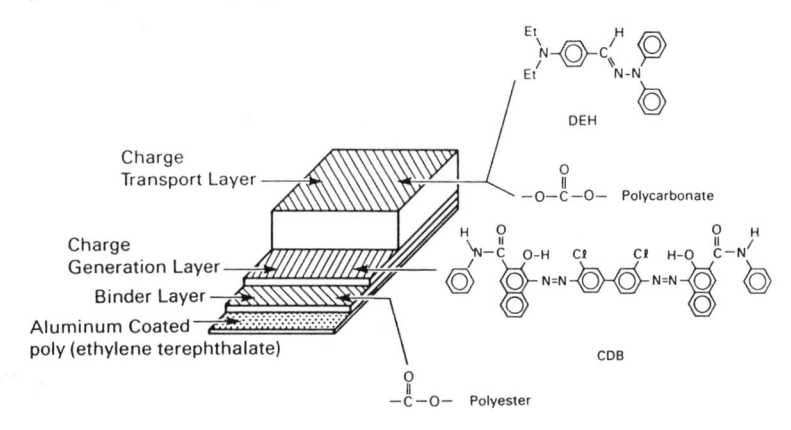

Figure 2. Composition of an organic layered photoconductor.

A summary of the types of molecules used by the industry has been presented by Nakanishi.[2] The general requirements for the hole transport molecule are a low ionization potential for efficient hole injection, and good chemical and physical stability of the hole, i.e. radical cation. A carrier generation dye should yield a large number of mobile carriers per unit absorbed photon under an applied electric field. Both of these materials should also have excellent coating characteristics and compatibility with polymers.

The function of each layer is illustrated in Fig. 3; negative ions are deposited on the surface of the photoconductor and create a potential of $\simeq$900 Volts. Upon exposure to long wavelength light, the strong electric field across the photoconductor ionizes the electronically excited dye molecule, chlorodiane blue. There is separation of the electron-hole pair, and the electrons move to the aluminum ground plane. The positive holes are injected into DEH at the interface between the charge generation and charge transport layers and, via a hopping conductivity mechanism,[8] are carried away from the interface to the surface of the photoconductor where they annihilate the negative surface charges. This entire process, the light-induced discharge, must be rapid in relation to the process time of a print;

for efficient throughput the transit time of the carriers should be of the order of $\simeq 100$ millisec. The drift mobility, being inversely proportional to the transit time, should therefore be $\simeq 10^{-6}$ cm^2/Volt.sec, or higher.

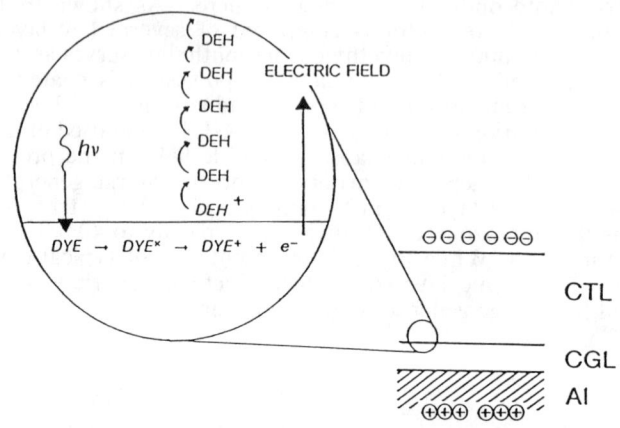

Figure 3. Illustration of the photodischarge process: carrier injection and hole transport.

2 OPTICAL ABSORPTION SPECTRA OF ORGANIC PHOTOCONDUCTORS

In the previous section, we have described some of the stringent requirements placed upon the photoconductor for efficient print throughput. This requires efficient electron-hole pair generation, efficient hole injection and rapid transport of holes via the DEH molecules. All of this is facilitated by chemical integrity of the photoconductor after fabrication in the various bulk layers and at the interfaces. Unfortunately, repeated exposure to light causes enough photochemistry to occur in these layers and interfaces that eventually destroys the photoconductor. For example, photochemically induced structural changes in a dye molecule like chlorodiane blue (CDB) may lead to poor carrier generation or injection efficiencies. Photochemistry in the hole transport layer may lead to charge trapping where current flow is otherwise expected. All of these effects would lead to a degradation of the initial electrical properties exhibited by the virgin photoconductor.

In this section, we discuss the effect of photochemistry of an organic photoconductor and introduce changes that are observed in its electrical properties. In Fig. 4, the optical absorption spectrum of the organic layered photoconductor is presented. The charge transport layer (CTL) absorbs light $\lambda < \simeq 470$ nm; hence, longer wavelength light is used to expose the CGL for carrier generation. Since the polymer binder in the charge transport layer does not absorb appreciably in the visible, all of the absorption is attributed to the hole transport molecule DEH. We note that the absorption tail is significant and DEH is readily accessible to photochemistry in this region. Fluorescent lights or white light discharge lamps have appreciable output in this spectral region. In the longer wavelength region, between 500 and 700 nm, the absorption bands are attributed to the dye chlorodiane blue. Chlorodiane blue is therefore sensitive to longer wavelength light, for example, a HeNe light source (630 nm) used

The Electrical Fatigue of Organic Photoconductors 199

to selectively "write," and equally, to fluorescent and white light discharge lamps. We first consider photoconductor fatigue caused by photochemistry of the hole transport molecule DEH.

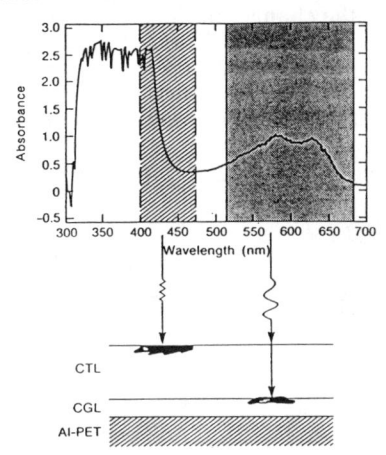

Figure 4. Optical absorption spectrum of photoconductor.

3 CHANGES IN ELECTRICAL PROPERTIES FROM DEH PHOTOCHEMISTRY

Upon absorption of light, DEH undergoes a photochemically induced cyclization reaction to an indazole derivative, 1-phenyl-3-(4-diethylamino-1-phenyl)-1,2-indazole, as shown in the scheme above. Associated with the photocyclization of DEH to the indazole derivative is an increase in residual voltage of the photoconductor. Emission and attenuated total reflection (ATR) infrared spectroscopy shows that the indazole derivative forms initially at the surface of the charge transport layer and, with continued exposure to light, forms deeper into the charge transport layer. The photochemistry is readily monitored by optical absorption studies, as shown in Fig. 5. DEH has an absorption maximum in the visible at 370 nm while the 265 nm absorption band characterizes the indazole derivative. As shown in Fig. 5, the 370 nm band decays as a function of incident light energy concomitant with evolution of a new absorption band at 265 nm. The photochemistry proceeds with a quantum efficiency of 0.4, and the changes in the optical absorption spectrum reveals an isosbestic point at 300 nm, indicative that the photochemistry of DEH results primarily in a single photoproduct, the indazole derivative, at least up to exposures of 26 J/cm^2. The indazole derivative has been isolated and completely characterized by x-ray crystal structure determination and

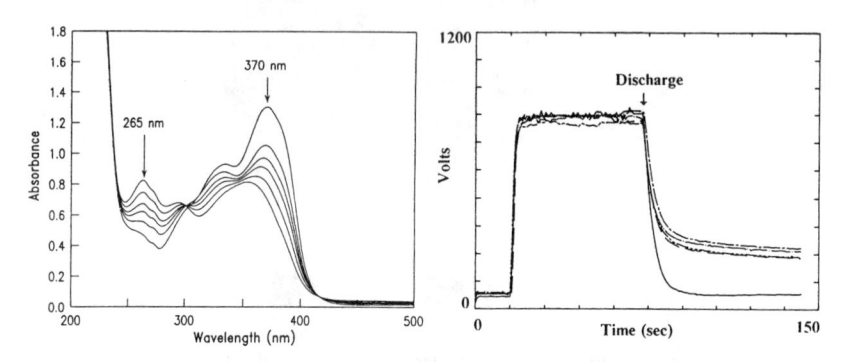

Figure 5 (left). Optical absorption spectra of 40% DEH in polycarbonate as a function of incident light (400 - 480 nm): 0, 3, 7, 10, 16 and 26 J/cm^2.

Figure 6 (right). Changes in the light decay electrical properties of a photoconductor, composed of a charge transport layer of 40% DEH in polycarbonate, exposed to light (400 - 480 nm): 0, 1, 2, 4 and 16 J/cm^2.

standard spectroscopic methods.[9] When the electrical properties of the organic photoconductor are measured simultaneously with the photochemical changes observed in the charge transport layer, an increase in residual voltage is observed, Fig. 6, during light decay. The increase in residual voltage is correlated with the formation of the indazole derivative in the charge transport layer. Less than 1 J/cm^2 of incident energy is apparently all that is required to cause enough photochemistry to occur in the charge transport layer to cause significant residual voltage. The residual voltage persists even in the presence of a discharge lamp, unacceptable in electrophotographic applications; thus the photoconductor is considered to be electrically fatigued.

The origin for the evolution of residual voltage is due to the presence of the indazole derivative in the charge transport layer. The indazole derivative acts as a barrier to hole transport because of its higher ionization potential.[3]

The initial electrical properties of the photoconductor may be recovered by annealing above the glass transition temperature of the charge transport layer ($\simeq$55°C). By annealing the photoconductor, DEH and the indazole derivative diffuse and mix in the charge transport layer. This acts to redistribute the indazole derivative throughout the bulk, thereby removing to some extent the barrier layer to hole transport. The number of times annealing may be used to recover the initial electrical properties is finite because DEH is depleted irreversibly.[3]

4 COMPUTATIONAL CHEMISTRY ASSISTS RESOLUTION OF FATIGUE

In the previous section, it was clearly demonstrated that the photochemical conversion of DEH, a hole transport molecule, to an indazole derivative, was correlated with electrical fatigue of the photoconductor. The crystal

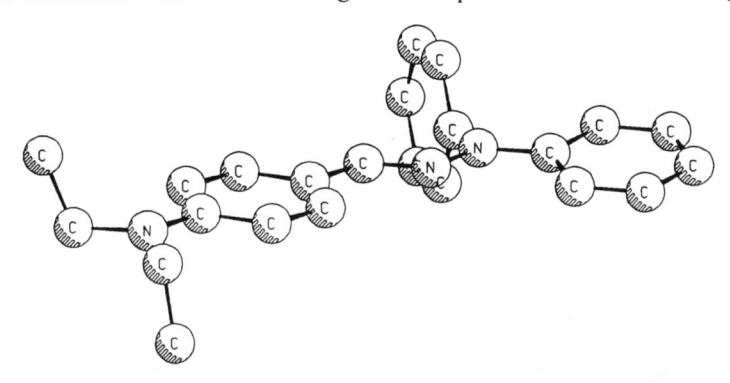

structure for DEH, shown above, reveals an extended π plane from the aniline amine nitrogen atom to one of the two hydrazone amine phenyl rings; the other hydrazone amine phenyl ring is essentially perpendicular to the extended π plane. The phenyl ring that is involved in the photocyclization is most likely the one that is perpendicular to the extended π plane. The postulated photocyclization reaction involves coupling of an ortho position of the hydrazone perpendicular phenyl ring with the imine carbon atom.[9] One reasonable manufacturing resolution to alleviate the photochemically induced electrical fatigue is to chemically modify the DEH molecule to the extent that the photocyclization reaction is inhibited, but simultaneously, its excellent hole transport properties and its excellent coating characteristics and compatibility in polymers are not altered. This cannot be decided upon decisively because the function of DEH is to transport holes, and it is precisely not known *a priori* where the hole is located on this rather large molecule. We are able to address these aspects using computational chemistry to obtain an understanding of where the hole might be located on the DEH molecule. This knowledge would give us insight into which portions of the DEH molecule are important in hole transport, and which might be important in the photocyclization reaction. Additionally, computational chemistry may provide new directions for synthesizing superior transport molecules.

A quantum chemical study using standard SCF calculations for the ground electronic states of neutral and positively charged (i.e. hole) DEH was used. The pertinent information extracted from the computational results are orbital plots for the highest occupied molecular orbitals (HOMO). As a first approximation, the electronic distribution given by the HOMO provides a description for those electrons of chemical relevance. The results of the computations are summarized in Fig. 7. In these figures, results from single point calculations using the minimal STO-3G basis set are shown, using a crystal geometry for DEH; thus no geometry optimizations are performed here. These calculations were performed on an IBM RISC 6000 Model 530 workstation, using the Gaussian 88 computer code,[10] and required $\simeq$45 min to complete per calculation. Computations for the DEH radical cation reveal that the hole may be localized on either the aniline amine nitrogen atom or the hydrazone amine nitrogen atom, since the difference in total energies is only 0.01 eV which is thermally accessible at room temperature. This is not surprising because lone pairs of electrons reside at these sites and are easiest to ionize. An additional state

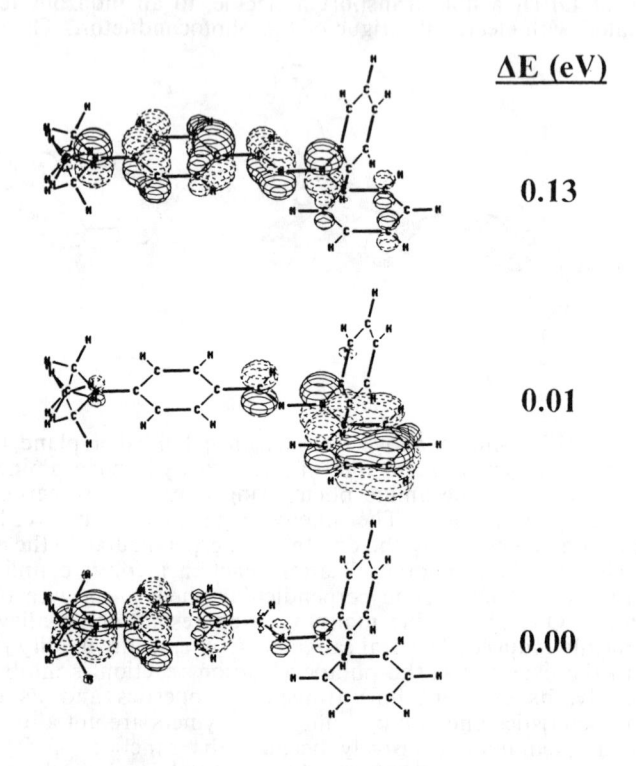

Figure 7. ROHF/STO-3G orbital plots for the DEH radical cation HOMO. Three different states and their relative energies are shown.

0.13 eV higher in energy is found for the delocalized hole. Here the charge is distributed over the entire DEH molecule that makes up the extended π plane. The important finding here is that whether the hole is localized on one or the other of the amine nitrogen atoms, or whether delocalized over the extended π plane, the perpendicular phenyl ring bonded to the hydrazone amine nitrogen atom does not contribute to the HOMO. Therefore the results of the calculations suggest that we may chemically modify or even replace the perpendicular phenyl ring, since it does not contribute electronically to the HOMO of the hole. Such modifications are therefore not expected to effect charge injection in DEH.

5 CHEMICAL MODIFICATION OF DEH TO INHIBIT ELECTRICAL FATIGUE

The results of the computations on DEH suggest that in order to inhibit the photocyclization of DEH to indazole, the chemical structure of DEH may be altered. The calculations indicate that the perpendicular phenyl ring may be chemically modified or replaced without greatly affecting the electronic structure of the hole. Using these results as a guide in the laboratory, a variety of DEH derivatives were prepared[6] to inhibit the photocyclization while simultaneously maintaining the low ionization potential of the parent DEH, important for hole injection, and the excellent processability and coating characteristics of DEH. Chemical structures

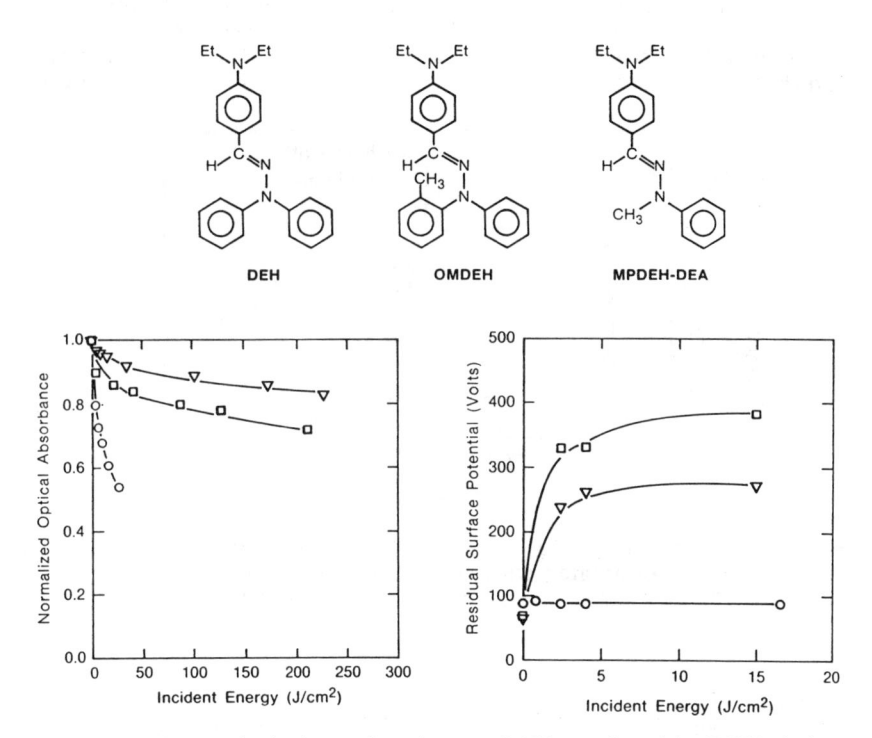

Figure 8. (left) Optical absorption decay of 370 nm band in DEH derivatives: DEH (○), OMDEH (□) and MPDEH-DEA (▽). The films are 40% by weight of the DEH derivatives in polycarbonate.

Figure 9. (right) Evolution of residual voltage for photoconductors exposed to light (400 - 480 nm). The hole transport layers are 40% by weight of DEH derivatives in polycarbonate: DEH (□), OMDEH (▽) and MPDEH-DEA (○).

of three such derivatives are shown above. The effects on the photochemistry and electrical properties of the photoconductors are summarized in Figs. 8, 9. As shown in Fig. 8, the photochemistry of DEH is considerably reduced when a methyl substituent is bonded to one of the ortho carbon atoms of the perpendicular phenyl ring, i.e., OMDEH. When the perpendicular phenyl ring is replaced with a methyl group, as in MPDEH-DEA, the photochemistry is further inhibited. The corresponding photo-induced discharge curves, Fig. 9, present parallel results. The residual voltage observed for photoconductors derived from OMDEH is roughly one-half as for the parent DEH, and no residual voltage is observed using MPDEH-DEA in the hole transport layer.

6 ELECTRICAL FATIGUE INITIATED BY CHLORODIANE BLUE

In the previous sections, electrical fatigue attributed to photochemistry of the hole transport molecule was discussed. Here we discuss the effect of exposing the carrier generation dye chlorodiane blue to long wavelength light. Changes in the electrical properties of the photoconductor can be correlated to changes in the optical absorption spectrum of the dye chlorodiane blue, when exposed to long wavelength light (550 - 800 nm), and the photo-induced discharge curves as a function of incident energy are

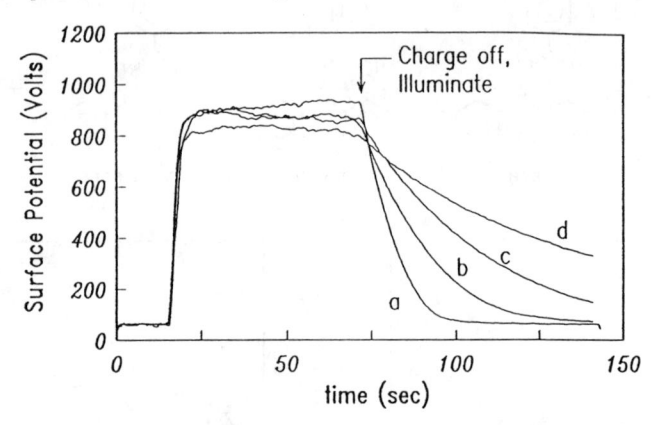

Figure 10. Light decay curves as a function of incident energy for a photoconductor formulated with chlorodiane blue and DEH in polyester: (a) 0, (b) 48, (c) 90 and (d) 120 J/cm^2. The polyester acid content is 4 μeq/g).

shown in Fig. 10. The light decay curves are observed to change significantly as a function of incident energy. After 48 J/cm^2 are incident on the photoconductor, photodischarge slows appreciably and with continued exposure approaches dark decay characteristics. Thus the photoconductor is no longer electrically conductive in the presence of a discharge lamp under an applied field. Concomitant with the changes in the electrical properties of the photoconductor are changes in the optical absorption spectrum of the dye chlorodiane, and these aspects are discussed next.

When solid state chlorodiane blue (CDB) is exposed to long-wavelength light, with no overcoating, i.e., no hole transport layer, the optical absorption spectrum of chlorodiane blue does not change as a function of incident energy, up to $\simeq$200 J/cm^2 at 102°C. With a polymer overcoat typically used to bind the hole transport layer, for example bisphenol-A polycarbonate or polyester coated atop chlorodiane blue, the optical absorption spectrum of the dye remains unchanged as a function of long wavelength light exposure, up to $\simeq$200 J/cm^2 at 102°C (Fig. 11). However, as soon as DEH is molecularly dispersed in the polymer overcoat, such as for a typical photoconductor formulation, for example, 40% DEH in polyester coated atop chlorodiane blue, Fig. 11, the optical absorption spectrum of the dye undergoes a decay or "bleaching" of its characteristic 630 nm absorption band that is readily evident after $\simeq$50 J/cm^2. The decay of the absorption band is accompanied by a color change of the dye from blue to copper. Moreover, whereas virgin chlorodiane blue is quite insoluble in common organic solvents, the copper colored photoproduct is readily

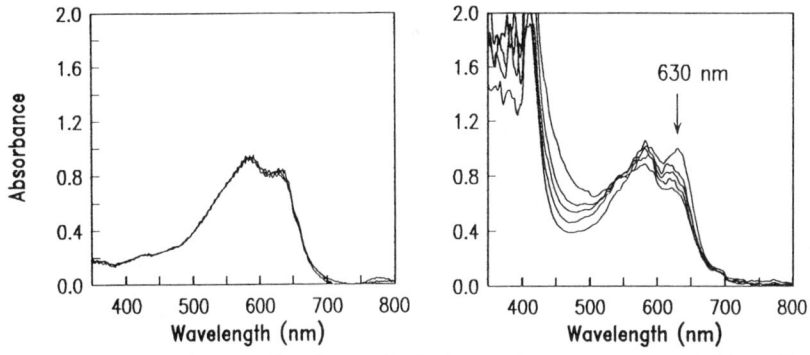

Figure 11. The changes in the optical absorption spectrum of chlorodiane blue. (left figure) A 20 μm thick polyester (acid content 54 μeq/g) overcoat as a function of incident energy (exposure to long wavelength light, 580 - 800 nm, at 102°C): 0, 92 and 184 J/cm^2. (right figure) A 20 μm thick overcoat of 40% DEH in polyester (acid content 4 μeq/g): 0, 48, 183, 412 and 641 J/cm^2.

soluble. This clearly illustrates that chlorodiane blue in the presence of DEH and long wavelength light, undergoes a photochemical reaction which transforms the dye and, for it to have become completely soluble, drastic changes in its chemical structure must have occurred. The result of the photochemistry is the consumption of chlorodiane blue and DEH at the carrier generation/hole transport interface, thus the two molecules responsible for carrier generation (CDB) and hole transport (DEH) are depleted at the interface, impeding photoconduction. Because this is a solid state photochemical reaction between CDB and DEH, diffusion and replenishing of the reactants at the interface are important and may limit the rate and extent of the photochemical reaction. For example, when the photochemical reaction is not allowed to proceed appreciably, e.g. ≃20 J/cm^2, annealing the photoconductor will recover its initial electrical properties.

The photochemically induced dye bleaching reaction of chlorodiane blue with DEH was monitored as a function of temperature, from which the activation energy for dye bleaching was determined. All experiments were conducted above the glass transition temperature of the hole transport layer (≃ 55 - 60°C) in a limited temperature range, 55 - 102°C, and initial decay rates were considered. Three polymers were chosen to formulate the hole transport layers with 40 weight % DEH: two polyesters that differ only in acid content, 4 and 54 μeq/g, respectively, and bisphenol A polycarbonate. The activation energies are 15, 19 and 20 kcal/mol for chlorodiane blue overcoated with 40% DEH in polyester (54 μeq/g acidity), polyester (4 μeq/g acidity), and bisphenol A polycarbonate, respectively. The activation energy is lowest when the photo-reaction between CDB and DEH occurs in a hole transport polymer host with higher acidity. We note also that the photochemically-induced dye bleaching reaction of chlorodiane blue is not limited to overcoats containing DEH; many hydrazone derivatives also induce photochemical dye bleaching. Additionally, squaraine dyes undergo a similar bleaching reaction in the presence of hole transport layer overcoats, producing similar changes in electrical properties as described above for chlorodiane blue.

7 COMPUTATIONAL CHEMISTRY AIDS INTERPRETATION OF INTERFACIAL DYE BLEACHING

In the previous section, experimental evidence was provided for electrical fatigue of organic photoconductors due to photochemistry initiated by the carrier generation dye chlorodiane. It was found that both chlorodiane blue and DEH were required at the interface to cause dye bleaching. Here we again invoke the assistance of computational chemistry to provide an understanding of the intermolecular orientation of CDB and DEH at the interface, and to provide a rationale for the photobleaching of chlorodiane blue.

The electrostatic potentials were calculated for DEH and both the azo-enol and hydrazone-quinone forms of one half of a chlorodiane blue molecule.[12] For DEH, a crystal geometry was used while for the two half-CDB molecules, they were optimized using SCF calculations at the 3-21G level of theory.[11] The results are illustrated in Figs. 12 and 13. Figure 12 contains the crystal structure of DEH and the optimized geometry of half of a chlorodiane blue molecule, necessary for identifying the individual atoms shown next in Fig. 13. For each of the three molecules isoenergy surfaces are shown (Fig. 13) at -25 kcal/mole. For DEH, the isoenergy surface includes the nitrogen lone pairs and the π electron clouds above and below the phenyl rings that constitute the extended π plane in DEH; note that the perpendicular phenyl ring is excluded. As the potential becomes increasingly negative, -37.5 kcal/mol, a minimum is located on the imine nitrogen atom. At similar potentials for chlorodiane blue, both in the azo-enol and hydrazone-quinone structures, the phenylamide group dominates the isoenergy surface at -25 kcal/mol. At -37.5 kcal/mol, minima are localized on the oxygen atom of the amide group. From these results, we might expect docking studies on the dye molecules to reveal that the amide groups dock on the positive areas of the potential surface on DEH. This appears to be a reasonable conclusion because during the photochemistry, the chlorodiane blue dye bleaches and the original color of the dye is imparted to a large degree by the phenyl amide group. In addition, docking studies may provide insight into charge injection and transport because of their dependance on intermolecular spacings.[13]

Preliminary docking studies of DEH on chlorodiane blue are reported using the Biosym Insight & Discover computer modelling software,[14] run on an IBM RISC 6000 Model 530 workstation. Both DEH and a hydrazone-quinone form of chlorodiane blue were optimized using the Biosym CVFF force field. In order to match the minimized geometries of these molecules using the Biosym CVFF force field to those previously calculated at the HF/3-21G level of theory,[11] suitable constraints were applied to the backbone dihedrals to insure planarity. For the docking studies, these same constraints were retained and additionally, HF/3-21G charges were used instead of the default charges provided in the Biosym CVFF force field. Docking studies were then performed by moving DEH relative to a stationary chlorodiane blue (CDB) and identifying the complexes with the lowest intermolecular energies (non-bonding and electrostatic energies). We were able to identify several low energy complexes in this manner by moving DEH about the stationary CDB. Several trends became evident. First, the complexes with the lowest energies prefer to have the π plane of DEH $\simeq$parallel to the π plane of CDB. Second, interaction of DEH with the Naphthol AS fragment[11] of chlorodiane blue provides lower energy complexes than those with the biphenyl moiety in

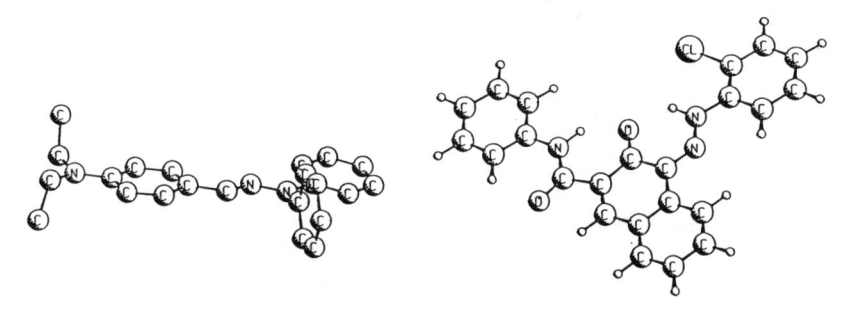

Figure 12. Crystal structure of DEH; optimized geometry of the hydrazone-quinone form of half of a chlorodiane blue molecule.

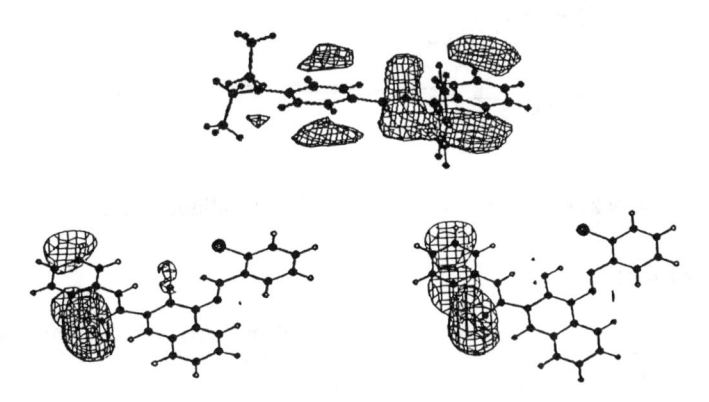

Figure 13. Isoenergy surface for DEH, hydrazone-quinone (left) and azo-enol (right) forms of half-CDB, at -25 kcal/mol.

CDB. Several such low energy complexes were identified and subsequently minimized. During the minimization, the individual molecules were still constrained to be planar; however, the complex was freely minimized. Additional minima for the CDB-DEH complex were located via molecular dynamics, followed by minimization. The dynamics runs were conducted at 900K for 250 ps (time step of 1 fs) in order to adequately sample the potential energy surface. The structures of the complexes were saved every 0.5 ps during the computer simulation. The resulting 500 dynamic structures were minimized and the low energy conformations examined. An illustration of the lowest energy complex found thus far, 156.8 kcal/mol, is presented in Fig. 14; intermolecular contacts $\leq$ 3.0 Å are represented by dashed lines. In this low energy complex, the positive hydrogen atoms on the diethyl groups of DEH "lock" onto the negative amido group of chlorodiane blue. Additional stabilization is provided for by the interaction of the perpendicular phenyl ring in DEH with chlorodiane blue. The distance between the two π planes is $\simeq$3.5 - 4 Å, as the two molecules are not exactly parallel. This brings the aniline portion of the DEH molecule directly above the Naphthol AS coupling fragment. The aniline amine nitrogen atom is one of two sites on DEH amenable to hole injection. More work is presently in progress.

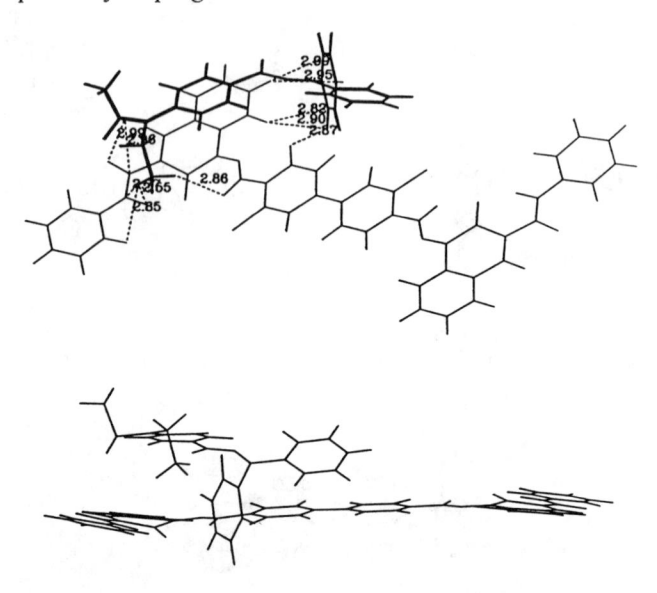

Figure 14. Docking of DEH on stationary chlorodiane blue.

REFERENCES

1. C. Carlson, 'Xerography and Related Processes,' J.H. Dessauer and H. Clark, eds., The Focal Press, New York, 1965, pp 15 - 49.
2. K. Nakanishi, Nikkei New Mater., 1987, 41.
3. J. Pacansky, R.J. Waltman, R. Grygier and R. Cox, Chem. Mater., 1991, 3, 454.
4. J. Pacansky, R.J. Waltman and R. Cox, Chem. Mater., 1991, 3, 903.
5. J. Pacansky and R.J. Waltman, Chem. Mater., 1991, 3, 912.
6. J. Pacansky, R.J. Waltman and R. Cox, Chem. Mater., 1991, 4, 401.
7. K.C. Nguyen and D.S. Weiss, Electrophotography, 1988, 27, 2.
8. J.X. Mack, L.B. Schein and A. Peled, Phys. Rev. B, 1989, 39, 7500.
9. J. Pacansky, H.C. Coufal and D.W. Brown, J. Photochem., 1987, 37, 293.
10. Gaussian 88: M.J. Frisch, M. Head-Gordon, H.B. Schlegal, K. Raghavachari, J.S. Binkley, C. Gonzalez, D.J. DeFrees, D.J. Fox, R.A. Whiteside, R. Seeger, C.F. Melius, J. Baker, L.R. Kahn, J.J.P. Stewart, E.M. Fluder, S. Topiol and J.A. Pople, Gaussian Inc., Pittsburgh, PA.
11. J. Pacansky and R.J. Waltman, J. Am. Chem. Soc., 1992, 114, 5813.
12. J. Pacansky, J. Carter, D. Jebens, R.J. Waltman and H. Seki, in 'Chemistry of Functional Dyes,' Vol. 2, Y. Shirota, ed., 1992, in press.
13. J.H. Slowik and I. Chen, J. Appl. Phys., 1983, 54, 4467.
14. P. Dauber-Osguthorpe, V.A. Roberts, D.J. Osguthorpe, J. Wolff, M. Genest and A.T. Hagler, Proteins: Struct., Funct., Genet., 1988, 4, 31.

The Chemistry and Applications of Oxonol–Iodonium Salts

Ranjan C. Patel[*] and Andrew W. Mott

MINNESOTA 3M RESEARCH LTD., THE PINNACLES, HARLOW, ESSEX CM19 5AE, UK

INTRODUCTION

New photo-labile molecules capable of response throughout the UV-VIS-Near IR range are of interest in the design of imaging media which can be addressed with emerging light sources. Diaryliodonium salts have found considerable application in radiation curable coatings[1,2], with the ability to be sensitised well into the visible region with dyes[2]. We describe in this paper a novel photo stimulated process involving oxonol dyes and iodonium cations (such as 1 and 2) with wide spectral response[3].

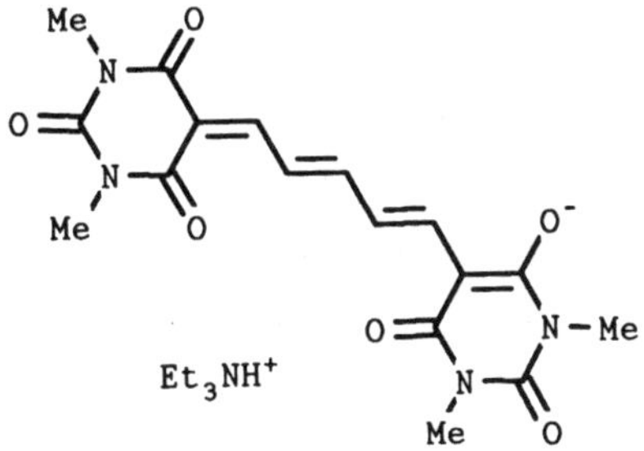

1

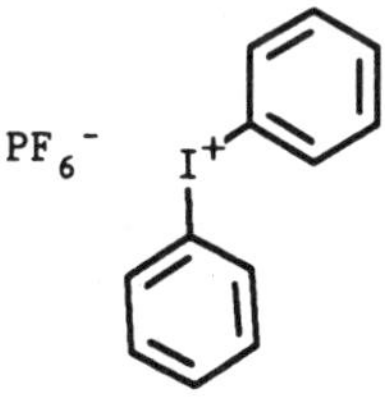

2

Figure 1: Structure of Oxonol Dye 1 and Diphenyliodonium Hexafluorophosphate 2.

EXPERIMENTAL SECTION

Oxonol dyes are prepared readily by condensing the appropriate keto-methylene (such as 3) with an anilino-polymethine precursor, such as 4. An example of a typical preparation is given below. A number of oxonol dyes were similarly prepared and are shown in Table 1, covering visible to near infra-red regions.

Oxonol Dye 1:

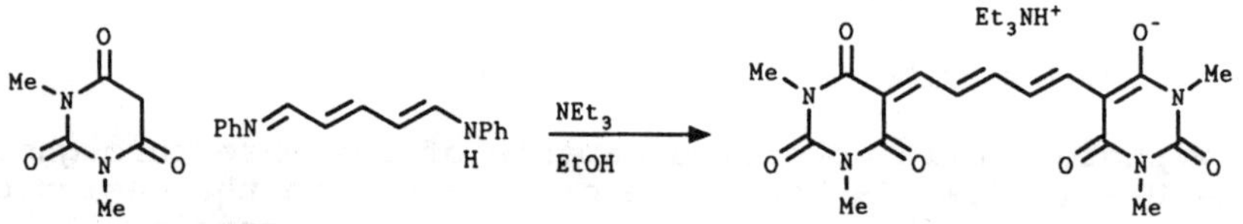

3   4   1

To a mixture of barbituric acid 3 (1.5g) and glutaconic aldehyde dianilide (1.4g) in ethanol (30g) was added triethylamine (1.4g). The resulting blue solution was refluxed for 0.5hr. After cooling to room temperature and addition of diethyl ether, a blue solid was filtered off and washed with copious ether. Blue needles were crystallised from ethanol, λ_{max} 590nm, ε_{max} 1.6X10^5 dm^3 mol^{-1} cm^{-1}.

Diphenyliodonium Hexafluorophosphate was prepared
according to the procedure of Beringer[4a].

Methoxyphenyl-phenyliodonium Trifluroacetate and 4-Butoxylphenyl-phenyliodonium Trifluoroacetate were
prepared following Beringer[4b]. These salts are soluble in water/ethanol/DMF solution and are useful in gelatin based coatings.

Oxonol-Iodonium Salts:

Attempts were made to prepare crystalline salts of the oxonol-iodonium by metathesis, using methanol-water mixtures in controlled lighting conditions. Very colourful viscous oils were isolated which resisted crystallisation. Because of the instability of these oils in light, most studies were subsequently made by using oxonol-ammonium salts, mixed with iodonium hexafluorophosphate such that in situ oxonol-iodonium salts formed.

Thin Film Coatings

Gelatin: The gelatin coatings were prepared by adding the dye as a solution in ethanol (0.05g in 2 ml) to a 8% solution of gelatin in H_2O at 40°C (10g). In controlled room lighting (e.g. red light for dye 1) then 4-methoxyphenyl-phenyliodonium trifluoroacetate (0.4g in 1 ml ethanol and 1 ml DMF) was added. The solution was coated using a wire wound coating bar (K Bar number 4, which deposits 36 micron thick wet coating) onto gelatin coated polyester film (100 micron), and dried in air. The film was handled in dim room light.

Butvar B-76: These solvent cast films were prepared by dissolving the oxonol dye (0.05g) in a 10g solution of polymeric binder Butvar B-76 (Monsanto, 10% in methyl ethyl ketone). Ethanol was added (1 ml). In controlled lighting (e.g. red for dye 1), diphenyliodonium hexafluorophosphate (0.1g) was added. The mixture was coated at KBar4 onto unsubbed polyester film (100 micron) and dried in air.

Oil Dispersed Coatings: A solution of oxonol dye 7 and 4-butoxyphenyl-phenyliodonium trifluoroacetate (0.1g) in ethanol (21 ml), DMF (2 ml) and di-n-butylphthalate (5.5g) was added dropwise to a solution of gelatin (16g) in distilled water (200g) held at 40°C and micronised with a Silverson dispersator. Green light was used during this formulation make-up. After dispersion, surfactant Tergitol TMN-10 (10%, aqueous, 6.5 ml) was added. The dispersion was coated onto subbed polyester in green light and dried in air. The coating was handled in dim green light.

Microdensitometer Testing: A modified microdensitometer was assembled, housing a Xenon Arc lamp (150W), to investigate the coated films using monochromatic irradiation. Figure 2 shows the layout of the test-rig. Narrow band filters (5nm width) were placed between the Xenon source (collimated to a 25mm^2 spot) and the test film sample, such that the response of the film with time of irradiation could be monitored, in terms of a plot of density vs time (secs). Typical power output through the narrow band filters was 2.4×10^{-3} W cm^{-2}, such that dye bleaching from a density of ca 2 required 120 secs or less. Figure 3B depicts a typical density time sequence at different wavelengths.

MOPAC Computational Calculations:

Calculations were performed using MOPAC version 5.0 (JJP Stewart MOPAC Ver 5.0 QCPE No. 455 (1989)) using the PM3 method running on a Silicon Graphics 4025 workstation using Sybyl version 5.1. The keyword PRECISE was used for all geometry optimisations. Excited state calculations were performed with C.I=4.

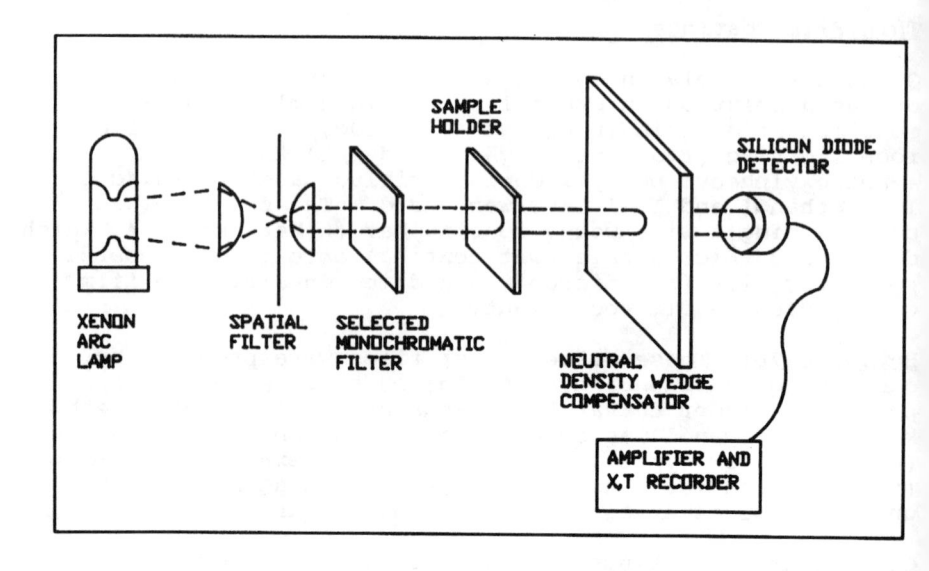

Figure 2: Density/Time Test-Rig for Spectral Response Analysis of Coated Films

THE OXIDATIVE IMAGING DYE BLEACH PROCESS (OXIM)

When a mixture of an oxonol dye (eg 1) and diphenyliodonium hexafluorophosphate (2) dissolved in ethyl methyl ketone is irradiated with light absorbed by the dye, rapid decoloration of the dye solution occurs (Fig. 3A). Thin film coatings of the dye-iodonium mixture in a suitable polymeric binder support (eg Butvar B-76) also showed this complete bleaching of the dye. Fig 3B depicts the response of oxonol dye 1 mixed with iodonium 2 when subjected to monochromatic irradiation, clearly showing the dependence on the dye absorption. Table 1 records a range of oxonol dyes which may be decolourised in the above manner.

A number of other dye classes were also investigated, such as cyanines and merocyanines. These dyes, however, in the present photo stimulated processes did not bleach well requiring excessive quantity of the iodonium salt and/or application of heat. The oxonol class is particularly well suited for dye bleach compositions with iodonium salts because of the possibility of dye iodonium$^+$ ion pair formation, enabling the close association necessary for efficient dye bleach. Ion pairing involving iodonium cations is well established, considering the work of Hacker et al on diphenyliodonium halides[5], the work of Neckers[6] on Rose Bengal onium salts and, more recently, the studies on diphenyliodonium anthracene-sulphonate salts[7].

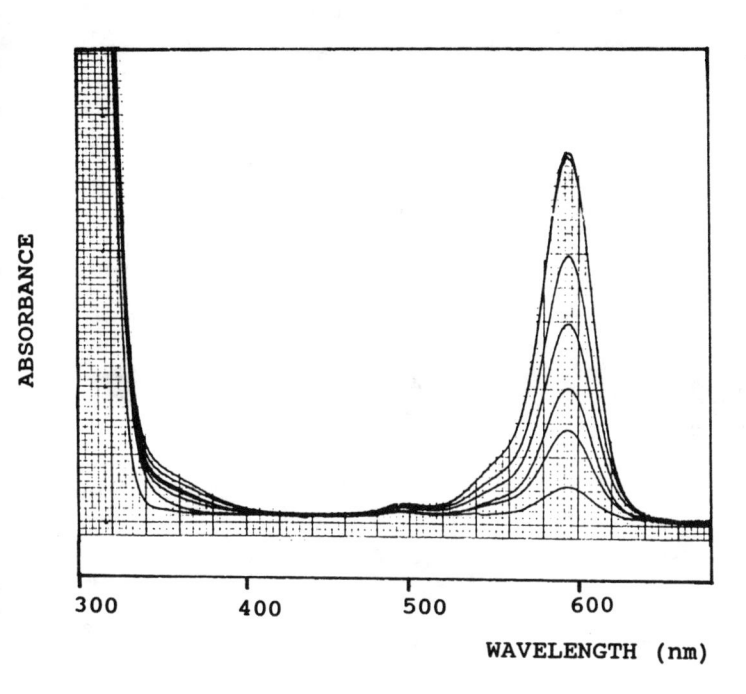

Figure 3A: Absorption Spectrum recorded at various time intervals after room light irradiation of a solution of oxonol dye 1 with excess diphenyliodonium hexafluorophosphate 2 in ethyl methyl ketone.

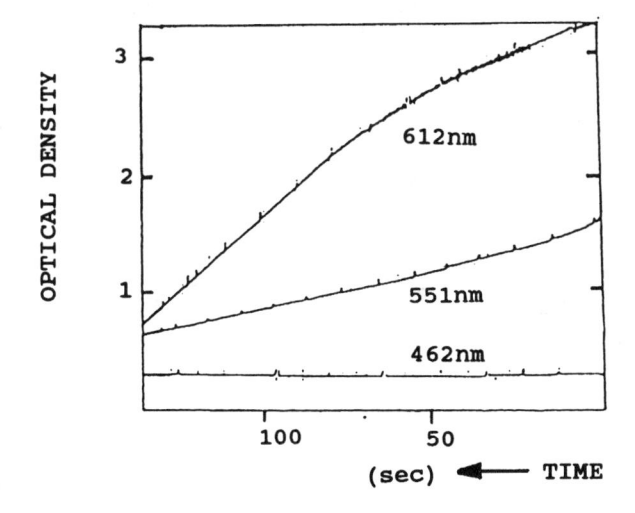

Figure 3B: Density/Time Response plots at various wavelengths for a coating of oxonol dye 1 with diphenyl iodonium hexafluorophosphate in Butvar B-76.

TABLE 1: Representative Oxonol Dyes Covering 360-790nm

Dye	Anion	λ_{max} nm (EtOH)	ε_{max}	Colour Change
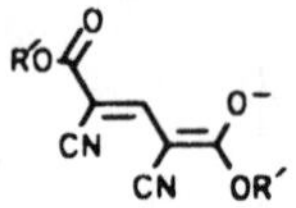	K$^+$	362	8×10^4	—
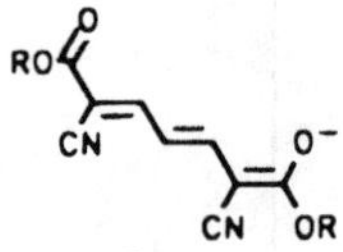	K$^+$	445	9×10^4	Yellow to Colourless
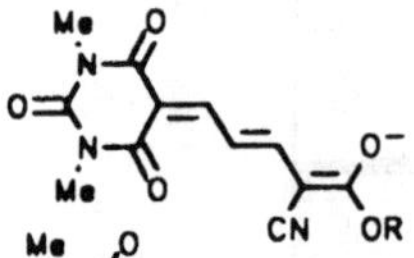	$^+$NHEt$_3$	457	1.1×10^5	Orange to Colourless
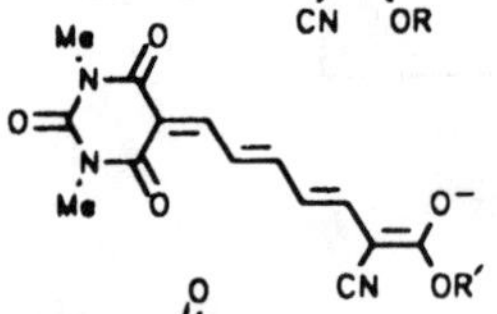	K$^+$	562	1.1×10^5	Magenta to Colourless
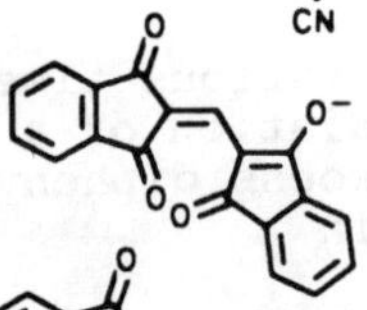	$^+$NHEt$_3$	453	1.0×10^5	No Change
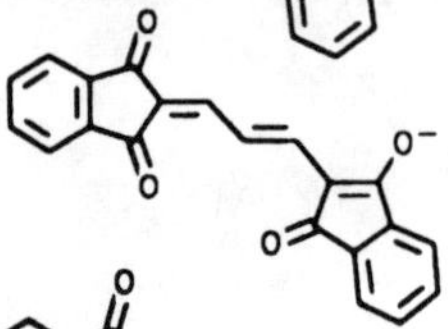	$^+$NHEt$_3$	555	1.5×10^5	Magenta to Pale Yellow
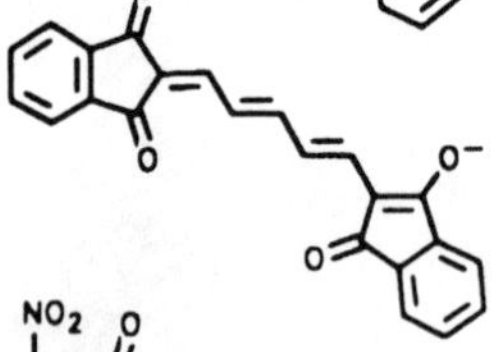	$^+$NHEt$_3$	655	1.5×10^5	Cyan to Pale Yellow
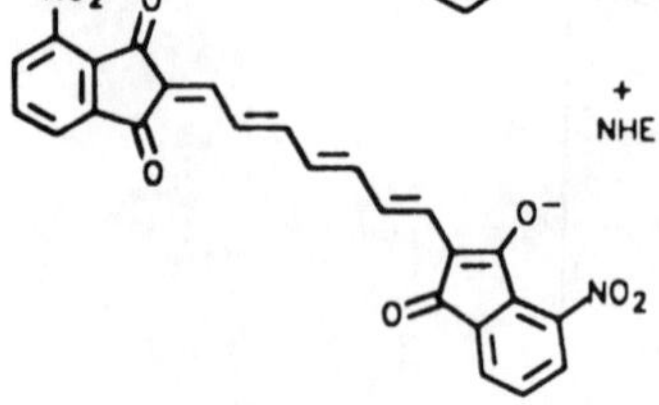	$^+$NHEt$_3$	785	1.3×10^5	Pale Blue to Pale Yellow

R = n-BUTYL

R' = ETHYL

The ease of ion-pairing between the anionic oxonol dye and the iodonium cation was influenced by solvent nature. Water or alcohol solutions of the oxonol-iodonium tended not to bleach well, nor did films of gelatin (cast out of water) containing the oxonol-iodonium. Solutions in ketone solvents, on the other hand, gave a faster bleaching response, as did films cast with Butvar B-76 out of ethyl methyl ketone. Figure 4 shows the typical curves obtained when comparing gelatin films, versus Butvar-B76. Best film response was, however, obtained from 1:1 oxonol- iodonium salts prepared in di-n-butylphthalate oil and this dispersed in gelatin, to form films in which the dye salts are dissolved and located in the micronised oil phase.

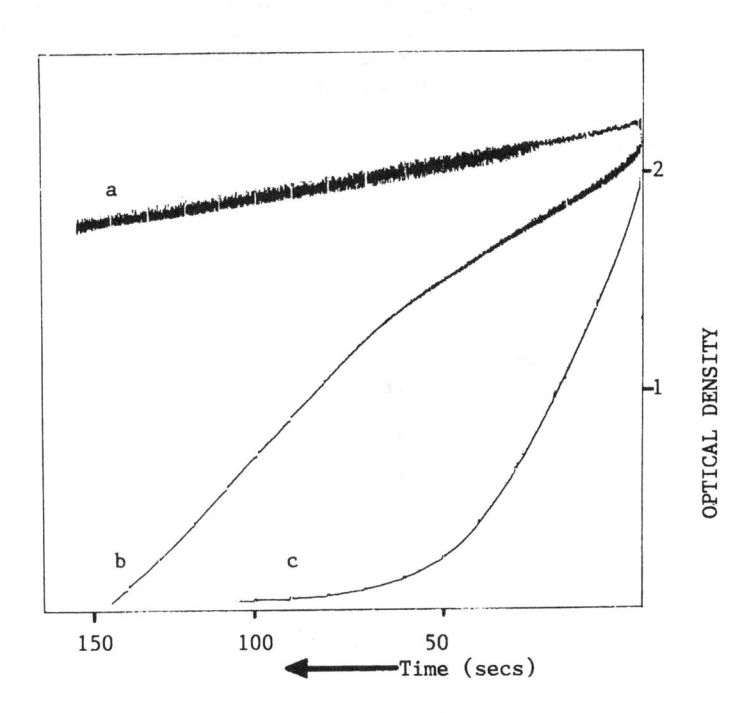

Figure 4: Comparison of Solvent/Binder conditions for Oxonol Dye 1, measured at 611 nm.

    a)   Gelatin film cast out of H_2O.

    b)   Butvar B-76 film cast out of MEK.

    c)   Di-n-butylphthalate dispersion of Oxonol 1 and 4-Butoxyphenyl-phenyl-iodonium salt in gelatin.

MECHANISTIC CONSIDERATIONS

The most likely mechanism is thought to involve electron transfer from the singlet excited state of the oxonol dye to the iodonium cation to produce the radical pair (5) (Figure 5). (It is to be noted that oxonol dye 1 fluoresces red in ketone solutions and that this fluorescence is quenched by iodonium salts).

The rapid fragmentation of the diphenyliodinyl radical aids forward momentum to result in the production of a phenyl radical in the vicinity of the dye radical; coupling of these two species results in dye bleach (Figure 5). This mechanism is not unreasonable in view of the observed solvent effect on the dye bleach process (described above) and the product distribution dependence on solvent nature recorded by Hacker et al[6] for iodonium halides. Further, Neckers[6] has put forward a similar singlet state oxidation of Rose Bengal by iodonium cations.

OXONOL[−]   DIPHENYL-IODONIUM[+]

hν

OXONOL•............DIPHENYLIODINYL•

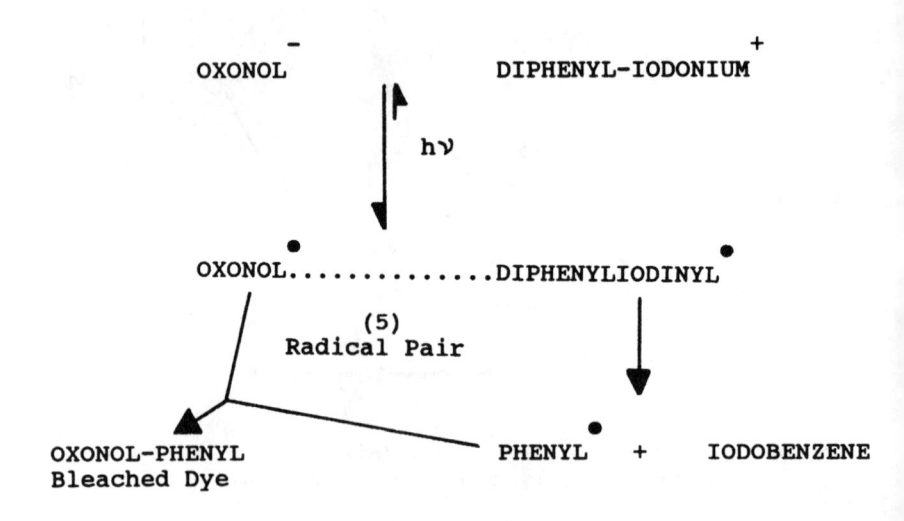

(5) Radical Pair

OXONOL-PHENYL   PHENYL• + IODOBENZENE
Bleached Dye

Figure 5: Mechanism Involving Intra Ion-Pair Electron Transfer

It is of interest that the converse mechanism has been suggested by Etter[8] for the photo-induced bleaching of cyanine dyes by organo-borate anions,[9] a process more extensively studied by Schuster et al[9] and applied for visible sensitisation of vinyl monomer polymerisation. Solvent studies of cyanine borates have showed that in nonpolar solvents the cyanine borate exists predominantly as ion pairs and that intra ion pair electron transfer occurs from the borate anion to the singlet cyanine dye on irradiation producing radical intermediates[10].

Phenylation Position

The recent study by Neckers[6] on the bleaching of Rose Bengal by iodonium ions describes the formation of O-phenylated products, some of which absorb in the visible. The oxonol dyes in the OXIM process particularly differ in that the bleached products absorb well below 400nm, typically at 360nm. C-Phenylation is inferred, particularly occurring at the ✳ position indicated below in Figure 6. Mono-O-phenylation would produce new dyes absorbing ca 400-450nm.

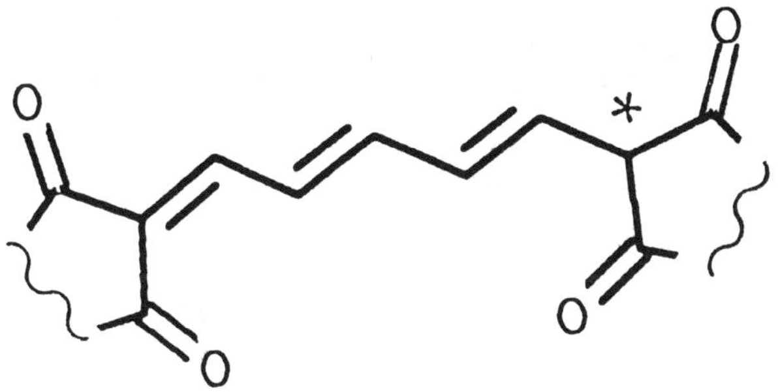

Figure 6: Position of Phenylation on the Oxonol Skeleton.

This position(✳) for phenylation is prone to steric effects from surrounding groups. A striking example of this effect is observed when comparing red absorbing oxonols 6 and 7 (Table 2). Though structurally very similar with similar absorptions, dye 6 is very reactive and unstable in the presence of iodonium ions, whereas dye 7 is 7 times slower to bleach and in some circumstances requires an induction period (see Figure 7).

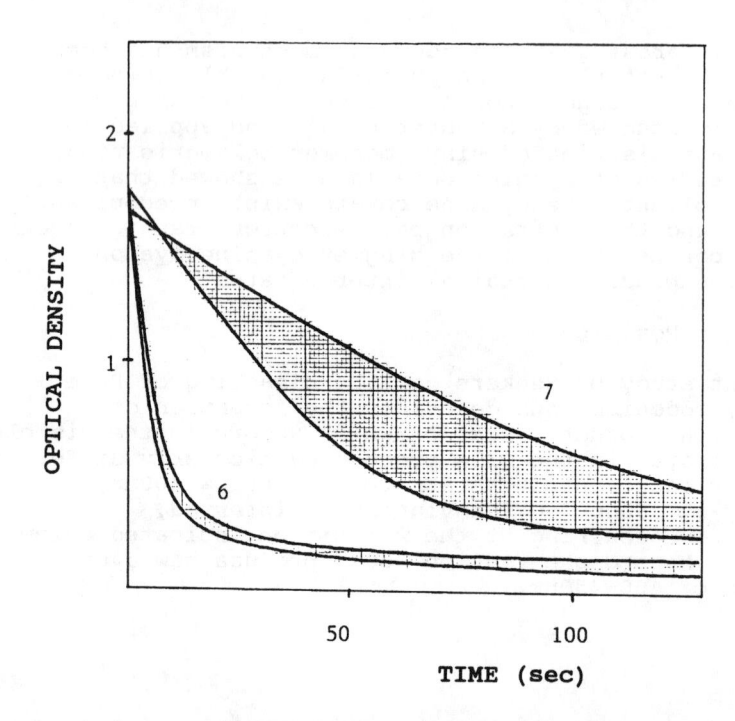

Figure 7: Plot of Density versus Time for Oxonol Dyes 6 and 7 measured at 671nm, coated as dispersions in di-n-butylphthalate in gelatin Two curves are shown for each dye: top curve measured at 25°C and lower curve at 60°C.

Table 2: Comparison of Red Absorbing Dyes 6 and 7

Dye	Structure	λ max (nm)	$t_{\Delta D=1}$ (secs)[a]	Sensitivity (J cm^{-2})[b]
6		662	7	0.017
7		650	50	0.120

a. Measured at 671nm. Time taken to bleach to an optical density of 1.0.

b. Calculated from energy flux through 671nm filter: 2.4×10^{-3} J sec^{-1} cm^{-2} x $t_{\Delta D=1}$ sec at 60°C.

MOPAC CALCULATIONS ON DYE STRUCTURES

The ground state ionisation potentials of dyes 6 and 7 were calculated as 4.24 and 4.89 eV respectively. These energies (407 and 470 kJ mol^{-1}) are too high for the dye to give up an electron to the iodonium from its ground state. Following excitation the calculated ionisation potentials of dyes 6 and 7 fall to 0.22 eV and 0.44 eV respectively. It is now much easier for the dye to lose an electron and transfer it to the iodonium. However, dye 7 appears to be still more resistive to electron transfer (greater ionisation potential) than does dye 6. This is consistent with the observed 10 fold decrease in sensitivity with this dye.

Following transfer of an electron to the iodonium it breaks down to the phenyl radical and phenyl iodide. This phenyl radical is in the vicinity of the dye radical and reacts with it to bleach the dye. Calculations have been performed with a starting geometry containing a phenyl radical positioned 5.31Å away from the oxonol radical. [Ph• to C=O 5.31Å, Ph• to CH 5.42Å, Ph• to C 5.35Å].

Following a geometry optimisation the phenyl group was found to bond preferentially to the carbon * of the dye (Figure 8), in both cases.

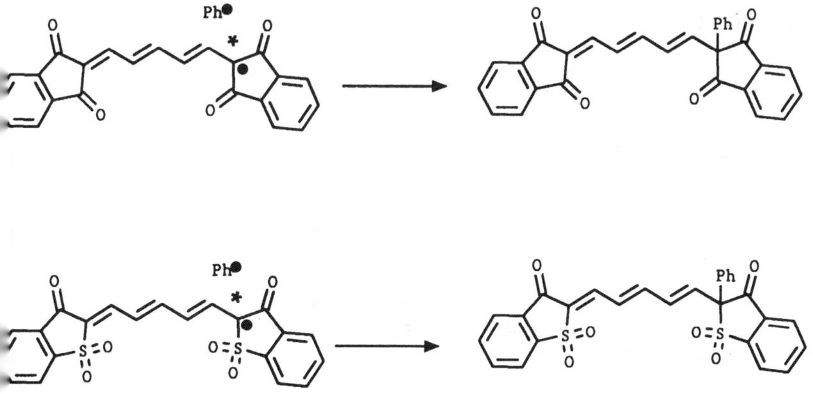

Figure 8

APPLICATIONS

The oxonol-iodonium photochemistry can be applied extensively because of the width of spectral response possible, by tailoring the dye absorption. Direct positive, full colour images may be produced by subtractive imaging of a combination of yellow, magenta and cyan dyes. tailoring the dye structure such that the iodonium bleaching of the yellow, magenta and cyan dyes is matched in sensitivity, a combination of 4 oxonols was obtained which demonstrated good colour separation yielding full colour images[3].

In addition to colour production, the oxonol-iodonium chemistry is capable of triggering polymerisation; in this case the phenyl radical, produced by dye photoreduction of the iodonium, is diverted to trigger polymerisation of incorporated, ethylenic monomers[11].

ACKNOWLEDGEMENTS

RCP thanks H.J. Pennicott, M.G. Fisher, J.H.A. Stibbard, R.G. Tye, R.W.C. Golder and D. Newman for their dedicated help and interest in the applications of the oxonol-iodonium salts. We thank the following officers of 3M who sponsored the OXIM work at various stages: I.J. Ferguson, A.N. Ferguson, G.F. Duffin, D.H. Dybvig, P.J. Finn and S.P. Birkeland. Prof. A.R. Katritzky is thanked for helpful guidance during the course of the project. We thank Mrs V.A. Morgan for typing this manuscript and finally, but not least, Dr. R.J.D. Nairne for a useful critique.

REFERENCES

1. J.V. Crivello and L.H.W. Lam, Macromolecules, 1977, 10, 1307.

2. G.H. Smith, Belgian Patent, 1975, 828,841.

3. R.C. Patel, H.J. Pennicott, I.J. Ferguson, US Patents, 1987 USP 4,701,402; USP 4,632,895; USP 4,769,459

4a F.M. Beringer, R.A. Falk, M. Karniol, I. Lillien, G.Masulto, M. Mausner and E. Somner, J.Am.Chem.Soc., 1959, 81, 342.

4b F.M. Beringer, H.E. Bachofner, R.A. Falk and M. Leff, J.Am.Chem.Soc., 1958, 80, 4279.

5. N.P. Hacker, D.V. Leff and J.L. Dektar, J.Org.Chem, 1991, 56, 2280.

6. D.C. Neckers and S.M. Linden, J.Am.Chem.Soc., 1988, 110, 1257.

7. K. Naitoh, T. Yamaoka and Umehara, Chem.Lett., 1991, 1869.

8. M.C. Etter, B.N. Holmes, R.B. Kress and G. Filipovick Isr. J. Chem, 1985, 25, 264.

9. S. Chatterjee, P. Gottschalk, P.D. Davies, M.E. Kurz, B. Sauerwein, X. Yang and G.B. Schuster, J.Am.Chem.Soc., 1990, 112, 6329.

10. G.B. Schuster, X. Yang, C. Zou and B. Sauerwein, J. Photochem. Photobiol.A: Chem, 1992, 65, 191.

11. R.C. Patel, T.V. Thien, D. Warner, European Patent Application, EPA 0444 7786A1, 1991.

Recent Developments in Photographic Development*

M. R. V. Sahyun

3M GRAPHIC RESEARCH LABORATORY, 3M CENTER, 201-3N-05,
ST. PAUL, MN 55144, USA

1 INTRODUCTION

The photochemistry of the silver halides is well-known and relatively well understood. The structure and photophysics of silver halide emulsion grains has been the subject of numerous reviews and monographs[1]. Massive photolysis decomposes them to metallic silver and free halogen. Exposure of a photographic coating, however, involves exposure of individual grains to, on average, 10-100 photons, to produce a latent image. It is generally understood that silver(0) clusters, as small as Ag_4, photochemically generated with high quantum efficiency, comprise the latent image. Quantum chemical calculations, using diatomics-in-molecules[2a] or extended Hückel formalisms[2b], as well as photoconductivity studies[3] indicate that such clusters on the surface of a silver halide grain are likely to be positively charged, i.e., Ag_n^+.

The latent image is rendered visible in the development step of the silver halide process. It is intended that the developer selectively reduces latent image-bearing grains completely to silver(0) and halide anion, whilst it, itself, is oxidized. The latent image thus functions as a nucleus on which the silver(0) phase grows. Photographic developers are solutions of reducing agents (usually organic), along with buffers, antioxidants (often sodium sulfite), and other addenda of specialized function. Developing agents have historically been classified according to the Kendall-Pelz rule[4]

$$A-(C=Y)_n-A$$

---

*) Contribution no. IIL-79a from the Imaging, Information and Electronics Sector Laboratories, 3M. Excerpted, in part, with permission from CHEMTECH, July 1992, 22, 418-424, copyright 1992, American Chemical Society.

where A is -OH, -NH$_2$, -NHR, or -NR$_2$, and Y is C or a heteroatom, usually N; n may be between 0 and 4, inclusively.

The chemistry of photographic development has been reviewed extensively[5-7]; the reviews cited do not, however, incorporate the latest thinking. It is still possible, however, to distinguish two limiting cases:
(a) <u>Direct</u> development in which the developing agent interacts with the surface of the silver halide grain to effect reduction of silver ions, presumably at the latent image/silver halide lattice interface; the electrochemistry of the silver-silver bromide electrode provides a model of this process[8].
(b) <u>Solution physical</u> development in which the silver ions are first dissolved out of the silver halide lattice by a complexing agent (which can be the sulfite anion of the developer antioxidant) and then redeposited in a redox reaction at the latent image site. Elegant stopped flow kinetic studies of this process have been carried out recently[9]. Solution physical development is especially important in so-called instant photography[10a] and in dry, thermally processed imaging media[10b].

Ideally, unexposed grains are unaffected; in reality, this is not the case. This unwanted process is as thermodynamically spontaneous as reduction of exposed grains, and leads to fog. Fog formation may be nucleated at silver(0) clusters formed thermally or by interaction of the silver halide with reducing impurities present in the emulsion and is termed <u>emulsion fog</u>. Alternatively, reduction of silver can be nucleated at reactive sites, e.g. defects, on the silver halide grain surface to produce <u>developer fog</u>. Heterogeneous attack of reducing agents on the silver halide grains at these sites may be compared to corrosion[23]; it tends to occur at grain edges and corners and is analogous to the well-known decoration of crystal defects by vapor deposition of metal[24]. Accordingly, emulsion fog and developer fog can be distinguished morphologically under the electron microscope[11].

2 AUTOCATALYTIC KINETICS

Early kinetic studies revealed that both direct[12,13] and solution physical[14] development are autocatalytic processes, in which the redox reaction occurs at the growing silver(0) surface. The course of an autocatalytic reaction is represented by the sigmoidal trace of Fig. 1. Typically, some time elapses before onset of visible image formation; this time, the <u>induction period</u>, t_i, may be measured as shown in the Figure. The chemistry occurring during this time is referred to as the <u>initiation phase</u> of development. Formation of a visible or measurable silver(0) deposit occurs during the <u>continuation phase</u>; kinetic measurements are normally possible only during this portion of the reaction, except with very sensitive instruments[12].

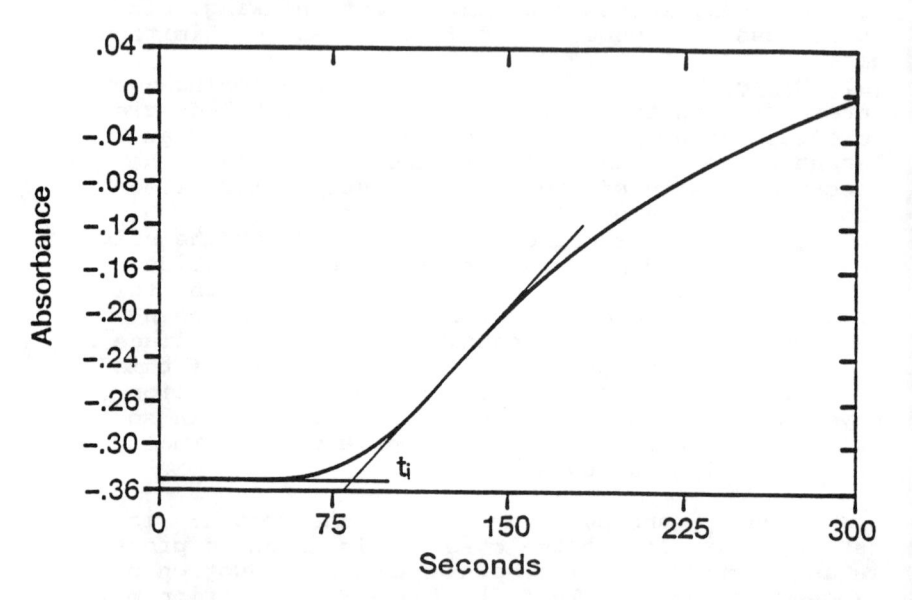

Fig. 1.-Growth of optical density, D, silver(0) in development of a liquid siver bromide emulsion with phenylhydrazine, as followed in the stopped-flow kinetic apparatus. (From Ref. 32).

The usual autocatalytic rate law is[15]:

$$1/(a_o - x_o) \cdot \ln[a_o(x_o + x)/x_o(a_o - x)] = kt \quad (1)$$

where a_o is the initial concentration of the limiting reagent (developing agent); x is the activity of the reaction product (silver(0)) at time, t; x_o is its initial activity. For a heterogeneous reaction, a_o is a surface concentration, proportional to, but not numerically equal to, the volume concentration of the developing agent in the developer according to an appropriate adsorption isotherm. In photographic practice, $x_o \ll a_o$, so that,

$$(1/a_o)\ln(x/x_o) = kt \quad (2)$$

a pseudo-first order form. From the slope of the plot of ln D, the developed optical density in the film or liquid emulsion, versus t, the autocatalytic rate constant, k, can be estimated; the intercept indicates relative activity of the silver(0) deposit present at the start of the autocatalytic process.

3 DEVELOPER SELECTIVITY

Malinowski pointed out in 1980 that formation of silver(0) from silver halide is a process of heterogeneous phase nucleation and growth[6a]. His model was subsequently elaborated by Moisar[16b]. Such phase formation is usually nucleated, as, for example, in the growth of crystals from a supersaturated solution. As noted above, the latent image centres serve as the nucleation sites. (Contrary to some suggestions[23b], there is no experimental justification to believe that photographic silver halide crystals are supersaturated in silver(0) prior to light exposure).

The developer solution may be considered as a redox buffer, i.e. a solution which tends to maintain a stable electrochemical potential even as its components react. Konstantinov and Malinowski[17] deposited silver(0) clusters of distributed size by vacuum evaporation onto an inert substrate and equilibrated them with various (Fe^{++}/Fe^{+++}) redox buffers. Not surprisingly they found that clusters smaller than some critical size tended to disappear while larger ones grew at their expense. They explained their results in terms of the excess surface free energy, ΔG_x, according to the Gibbs-Thomson equation

$$\Delta G_x = 2\sigma V_m/r \tag{3}$$

where σ is the specific surface energy of silver(0); V_m is the molal volume of silver(0). Accordingly, the electrochemical potential (in volts) of the developer with which the particle of radius, r, is in exact equilibrium will decrease with decreasing r as $F \cdot \Delta G_x$ (where F is the Faraday constant).

The relationship embodied in eq. 3 may not necessarily extend to latent image size clusters, given the lack of metallic character of small metal clusters and the tendency of molecular-sized clusters to exhibit size-quantized properties. As early as 1968, however, Trautweiler[18] carried out semi-empirical calculations which predicted monotonic increase in reduction potential of silver clusters, Ag_n, with increasing n, i.e. the behaviour observed by Malinowski extrapolated to the latent image size range. This trend has been confirmed experimentally for latent image centers in a real photographic emulsion[19], and for small silver clusters in aqueous solution by a series of elegant pulse radiolysis experiments[20,21].

An implication of this model is that developing agents which are stronger reducing agents ought to recognize smaller silver clusters, formed, in turn, by fewer photons, i.e. stronger developers enable higher photographic speeds from the same emulsion. In practice, however, these more active developers are also more prone to developer fog, which, in turn, limits the extent to which photographic speeds can be "pushed". If the latent image, in fact, has to be a cationic cluster, Ag_{n+1}^+, then the smallest possible value of n is that at which,

$$Ag_n + Ag_i^+ \rightleftharpoons Ag_{n+1}^+ \qquad (4)$$

where Ag_i^+ represents an interstitial silver ion in the silver halide lattice, is exoergic. The quantum chemical calculations previously cited[2] indicate that the threshold for thermodynamic stability of Ag_{n+1}^+ occurs at n = 4, in good agreement with photographic experiments in which cationic silver clusters, formed in an ion beam apparatus, were size selected in a mass spectrometer and deposited on emulsion grains, which were then tested for developability[22].

4 INITIATION AND CONTINUATION OF DEVELOPMENT

As noted above, an induction period elapses between beginning of the development reaction and the first appearance of a photgraphic image. James[25] reported that the duration of the induction period was unexpectedly related to the charge on the developing agent, as shown in Fig. 2.

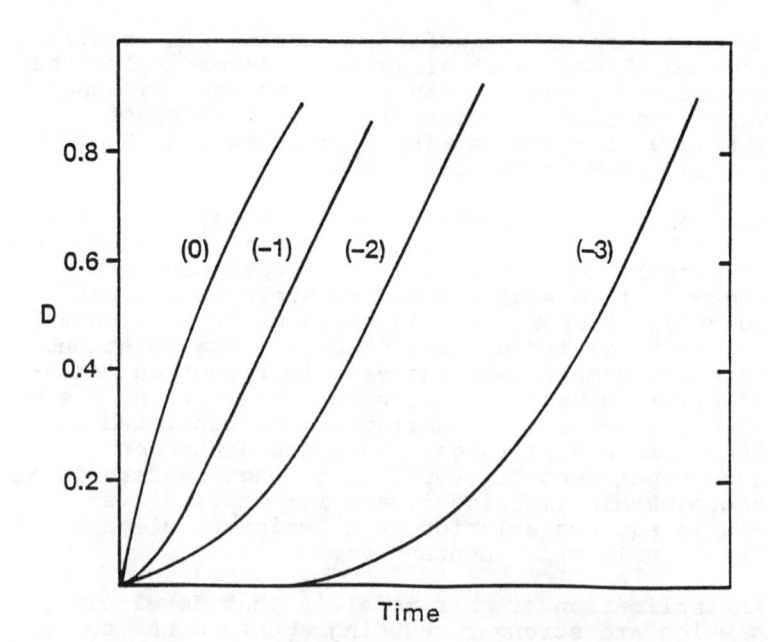

Fig. 2.-Charge barrier effect: development induction period increases with increasing negative charge on the developing agent. (From Ref. 25a).

Under photographic conditions, silver halide grains usually bear a negative surface charge; electrostatic repulsion thus decreases concentration of developing agent at the silver halide grain surface, a_o, with increasing negative charge on the developing agent molecule. This explanation implies that the developing agent is adsorbed to the silver halide surface during the initiation stage, but not necessarily during the continuation stage, of development; accordingly the chemical mechanisms of initiation and continuation may be different.

Additional evidence that initiation and continuation stages are mechanistically distinct came from our Laboratories, where we showed[26] that antifoggants, e.g. benzotriazole or 5-nitrobenzimidazole, inhibit the initiation stage, but not the continuation stage, of development. These compounds, added to practical developers to control developer fog, chemisorb at reactive sites on the silver halide grain surface[27], presumably in competition with the developing agents.

Sequential operation of different mechanisms may also apply to solution physical development. Results of a stopped-flow kinetic study[28] in which silver ions in solution (as $AgBF_4$) oxidized a p-phenylenediamine derivative colour developing agent (PPD) with concomitant dye formation, indicates existence of two mechanistic regimes, as shown in Fig. 3. Initially reaction is first order in both PPD, the limiting reagent, and Ag^+, and corresponds to one-electron oxidation of PPD in the rate limiting step. Subsequently an autocatalytic stage may be observed in which rate of dye formation is independent of $[Ag^+]$, as expected from eq. 2. Stoichiometry of this reaction corresponds to the usual two-electron oxidation of PPD; transfer of the second electron must be rate-determining under these conditions. The point of transition from pseudo-first order (initiation) to autocatalytic kinetics (continuation) depends, as shown, on $[Ag^+]$.

In addition to the kinetic distinction, a morphological distinction, in terms of the form of the silver(0) deposit, may exist as well between initiation and continuation stages of development. Under conditions of solution physical development, silver(0) is produced as a more-or-less spherical particle, one for each latent image or development centre. Under the electron microscope[13,29] it can be seen, however, that direct development also produces spherical silver(0) deposits during the initiation stage, but the continuation mechanism produces a filamentary mass in which the silver(0) deposit exhibits fractional dimensionality. With either 2-chloro-4-aminophenol or ascorbic acid as the developing agent, the transition apparently occurs when ca. 2 - 5% of the total silver by mass is reduced.

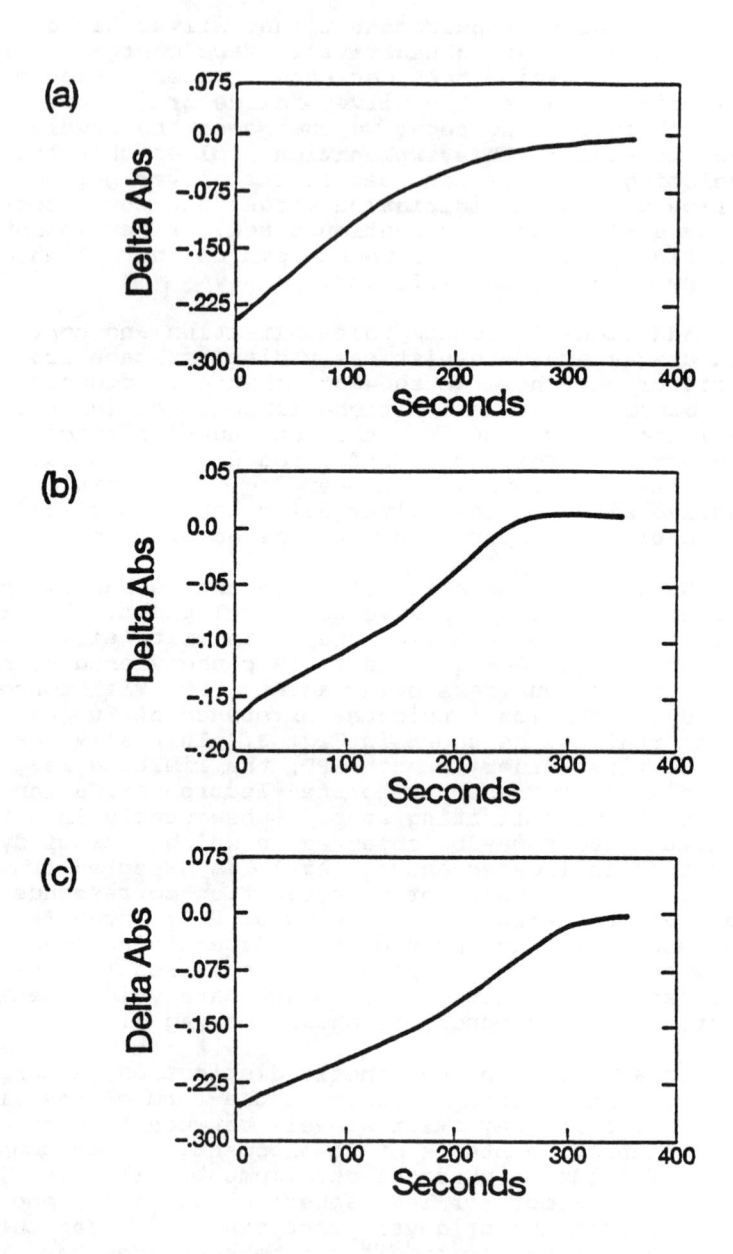

Fig. 3.-Growth of optical density owing to dye formation as observed in the stopped-flow apparatus for solution physical development with PPD (1.0×10^{-5} M) and (a) 0.10 M AgBF$_4$; (b) 0.025 M AgBF$_4$; (c) 0.0075 M AgBF$_4$. (From Ref. 28).

Pontius and Willis[13] have also shown that during direct development, the rate of reduction of silver during the continuation stage is: (a) proportional to the surface area of developed silver(0); and (b) limited by the rate of electron transfer from the solution to the silver(0) deposit. We subsequently confirmed these results for ascorbic acid development[15], and showed that this surface area dependence was the basis for validity of eqs. 1 and 2. The same considerations clearly apply to solution physical development with PPD.

Still additional evidence for mechanistic distinction between the initiation and continuation phases of development comes from linear free energy relationships obtaining during development. Dependence of k on the Gibbs free energy change occurring during that reaction, ΔG, is known as the Brønsted slope, β,

$$- k_b T \cdot d(\ln k)/d \Delta G = \beta \qquad (5)$$

Kochi and co-workers[30] demonstrated that β's of ca. 0.5 are typical of <u>outer sphere</u> electron transfer reactions, while β's approaching unity characterize <u>inner sphere</u> reactions. Earlier studies on the initiation stage of development with substituted p-phenylenediamines (commonly used in colour photographic development) and p-aminophenol reducing agents[31a] and with hydroquinone derivatives[31b] suggested β's of near 0.5. Stopped-flow kinetic studies of the reduction of <u>liquid</u> phtographic emulsions by substituted phenylhydrazines yielded β's between 0.38 and 0.62 [32]. The only well-controlled linear free energy correlation for continuation of direct development involved use of substituted p-phenylenediamines, and exhibited $\beta = 0.93$, indicative of inner sphere reaction[33]. This classic correlation is shown in Fig. 4. A stopped-flow study of solution physical development with p-phenylenediamine yielded a similar estimate of β [9d]. We have inferred[32], at least for development with p-phenylenediamines, that the transition from initiation to continuation stage also corresponds to transition from outer to inner sphere reaction.

Intercepts of plots according to eq. 2 for ascorbic acid development[15] suggest that x_o corresponds to a silver deposit whose activity is a few percent of that obtaining at the completion of development, x_{max}, i.e. a few percent of the total amount of silver reduced in the development reaction, consistent with the electron microscopy (above).

In the stopped-flow kinetic study of development of liquid AgBr emulsion by substituted phenylhydrazines[32], we pre-nucleated the emulsion with $NaBH_4$. With increasing $[BH_4^-]$, t_i decreased monotonically,

reaching zero when [BH$_4^-$] was equivalent to ca. 0.1 %
of the AgBr present. By inference at least this much
silver must be reduced by the mechanism of the initiation stage before transition to autocatalytic, continuation stage mechanism occurs.

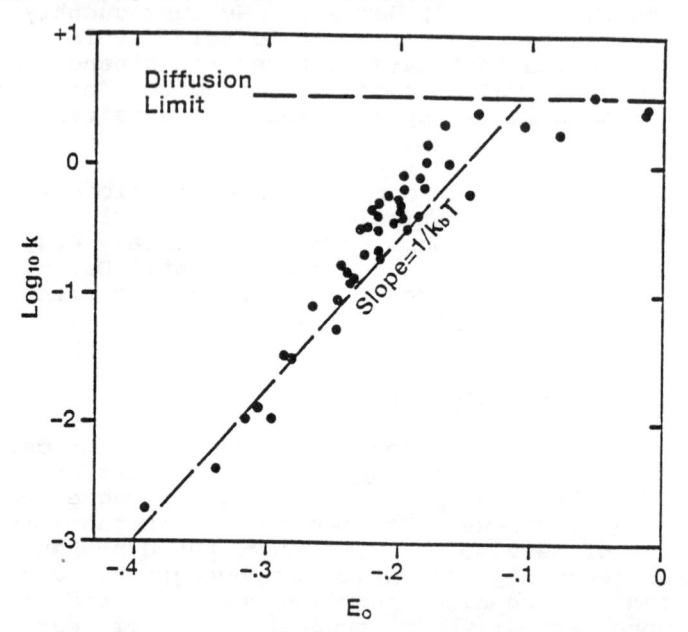

Fig. 4.-Linear free energy correlation of rate of
development of a silver halide film by substituted p-
phenylenediamine developing agents with developing
agent oxidation potential, -E$_o$. Maximum observable
rate is limited by diffusion. (From Ref. 33).

5 KINETICS OF PHASE FORMATION

The rate of electron transfer, rate constant k$_{et}$,
from the reducing agent to the silver cluster is rate
determining and proportional to the available
silver(0) surface area, S (vide supra). Thus

$$d[Ag^o]/dt = k_{et}K_{ads}S \quad (6)$$

where K$_{ads}$ is the equilibrium constant for adsorption
of developing agent to silver(0). This is, of course,
the simplest possible case, and more complex adsorption isotherms may obtain in the real world. The rate
constant, k$_{et}$, is determined by the electrochemical
energetics for reduction of silver, in accord with the
Marcus and Levich models of electrode reaction dynamics[34]. Then

$$-k_bT \ln K_{ads} = \Delta G_{ads} = (\Delta G^o_{ads} - \Delta G_x) \quad (7)$$

where ΔG_x has been defined in eq. 3 as $2\sigma V_m/r$. Substitution of eq. 3 into eq. 7 and eq. 7 into eq. 6 yields (if we assume spherical geometry for the growing silver deposit)

$$d[Ag^o]/dt = k_{et}(4\pi r^2)\exp[-(\Delta G^o_{ads} - 2\sigma V_m/r)/k_b T] \quad (8)$$

A plot of this equation, using Malinowski's value for σ[17] is shown in Fig. 5. Differentiation of eq. 8, as $d \ln\{d[Ag^o]/dt\}/dr$, and setting the derivative equal to zero, allows us to demonstrate that a minimum in the rate corresponds to the case

$$\Delta G_x = 2 k_b T = 2\sigma V_m/r_o \quad (9)$$

which, at room temperature, corresponds to $\Delta G_x = 0.052$ eV and $r_o = 20$ Å, or ca. 2000 silver atoms. We identify this minimum with the point of transition from initiation to continuation of development.

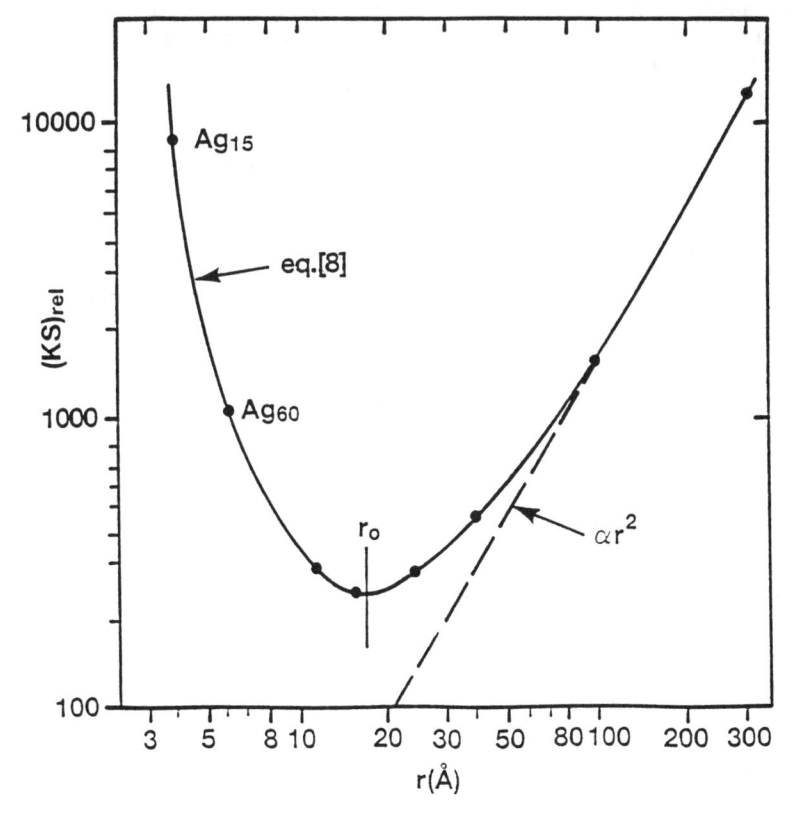

Fig. 5.-Plot of eq. 8; dependence of surface reactivity of a spherical silver(0) deposit on radius, r; linear dependence on surface area is approached asymptotically with increasing r; r_o marks the critical size for onset of autocatalytic growth.

In fact, the prefilamentary stage of direct development persists until the sperical deposits are about 10X greater in radius than estimated above, but the autocatalytic rate law may obtain during much of this time[13]. The discrepancy between this theoretical estimate and the apparently larger degrees of silver reduction required for onset of the continuation stage experimentally, e.g. in the borohydride pre-nucleation experiment, may be accounted for, in part, by multiple nucleation sites on individual grains.

6 DISCUSSION

We are now in a position to integrate these observations into an overall scheme of photographic development applying to both direct and solution physical development. Light exposure of silver halide grains in a photographic emulsion leads to formation of small silver(0) clusters, which are likely to be positively charged. The photographic developer acts as a redox buffer solution: clusters smaller than a critical size tend to disappear oxidatively, while those larger than the critical size grow during the initiation stage of development[16]. Two mechanisms have been proposed for this process[23]: (a) solution physical development; and (b) a zero-order direct transfer of electrons from the developing agent to silver halide grain, thence via the condution band of silver halide to the growing silver(0) center. Only (b) is consistent with the charge barrier effects, and can operate under conditions where only direct development is feasible.

Under these conditions, until the growing silver(0) deposit reaches a critical size (calculated above as about 2000 silver atoms), the slow, zero-order electron injection process allows it to grow. Interstitial silver ions from the silver halide lattice attach themselves to the cluster and are reduced by the injected electrons[8]. This process is analogous to that by which the latent image grows during light exposure. In this stage of development, oxidation of the developing agent appears to proceed by an outer sphere mechanism[32].

Only after the critical size is reached, can the silver(0) center can grow autocatalytically whether by direct or solution physical development. In direct development the process may be envisioned as a short-circuited electrochemical cell: oxidation of the developing agent takes place at the surface of the cluster in contact with the solution; reduction of silver ions takes place at its interface with the silver halide lattice. In either direct or solution physical development, autocatalysis results from reactivity increasing in proportion to silver(0) surface area. Electron transfer from developing agent strongly adsorbed (chemisorbed ?) to the metallic silver surface may then be an inner sphere reaction.

As a model of solution physical development, Henglein and co-workers[35] have studied spectrophotometrically reduction of $Ag(CN)_2^-$ by organic radicals at colloidal silver(0). They emphasize that this reaction, like development, does not simply involve electron transfer between Ag^+ and developing agent on the surface, but between developing agent and the whole silver cluster, to which soluble silver ion complex is adsorbed.

7 CONCLUSION

Primarily as a result of advances in fundamental sciences over the past decade, a good mechanistic understanding of the chemistry of photographic development is now in hand. This understanding can be put in quantitative terms to provide a mathematical model of the development stage of the silver halide process[36], which, in turn, becomes a useful tool for the engineering of photographic emulsion systems of improved performance.

REFERENCES

1. D. M. Sturmer and A. P. Marchetti, in Imaging Processes and Materials, 7th ed., J. Sturge, V. K. Walworth, and A. Shepp, Eds., Van Nostrand-Reinhold, New York, 1989; p. 71, and references cited therein.
2. (a) J. W. Mitchell, Photogr. Sci. Eng. 1978, 22, 1; (b) M. R. V. Sahyun, Photogr. Sci. Eng. 1978, 22, 317.
3. H. Hada, M. Kawasaki, and T. Yoshida, Photogr. Sci. Eng. 1981, 25, 201.
4. W. Pelz, Angew. Chem. 1954, 66, 231.
5. M. R. V. Sahyun, J. Chem. Educ. 1974, 51, 72; W. Jaenicke, Photogr. Sci. Eng. 1962, 6, 185.
6. W. Jaenicke, in Advances in Electrochemistry and Electrochemical Engineering, vol 10, H. Gerischer and C. W. Tobias, Eds., Wiley, New York, 1977.
7. M. R. V. Sahyun, Electrochim. Acta 1978, 23, 1145.
8. C. J. Battaglia, Photogr. Sci. Eng. 1970, 14, 275.
9. (a) U. Nickel, K. Kemnitz, and W. Jaenicke, J. Chem. Soc., Perkin II 1978, 1188; (b) U. Nickel, Ber. Bunsenges. Phys. Chem. 1981, 85, 266; (c) U. Nickel, in Progress in Basic Principles of Imaging Systems, F. Granzer and E. Moisar, Eds., Vieweg, Wiesbaden, 1986; p. 396; (d) U. Nickel, N. Rühl, and B. M. Zhou, Z.

Phys. Chem. 1986, 148, 33.
10. (a) V. K. Walworth and S. H. Mervis, in Ref. 1, p. 181; (b) D. H. Klosterboer, ibid., p. 279.
11. S.-X. Ji, P. Hu, and X.-M. Ren, in Ref. 9c, p. 402.
12. H. Frieser and M. Schlessinger, J. Photogr. Sci. 1972, 20, 192; J. Karrer and W. F. Berg, ibid. 1971, 19, 143.
13. R. B. Pontius and R. G. Willis, Photogr. Sci. Eng. 1973, 17, 21; 326.
14. T. H. James, J. Amer. Chem. Soc. 1939, 61, 648; 2379; R. B. Pontius and J. S. Thompson, Photogr. Sci. Eng. 1957, 1, 45; R. M. Cole and R. G. Pontius, ibid. 1961, 5, 154.
15. M. R. V. Sahyun, Photogr. Sci. Eng. 1974, 18, 504.
16. (a) J. Malinowski, Growth and Properties of Metal Clusters, J. Bourdon, Ed., Elsevier, Amsterdam, 1980; p. 303; (b) E. Moisar, Photogr. Sci. Eng. 1982, 26, 124.
17. I. Konstantinov and J. Malinowski, J. Photogr. Sci. 1975, 23, 1; 145.
18. F. Trautweiler, Photogr. Sci. Eng. 1968, 12, 138.
19. P. J. Hillson, J. Photogr. Sci. 1974, 22, 31.
20. (a) A. Henglein, in Modern Trends of Colloid Science in Chemistry and Biology, H.-F. Eicke, Ed., Birkhäuser, Basel, 1985; (b) P. Mulvaney, T. Linnert, and A. Henglein, in Symposium on Electronic and Ionic Properties of Silver Halides, B. Levy, Ed., IS&T, Springfield, VA, 1991, p. 239.
21. M. Mostafavi, J. L. Marignier, J. Amblard, and J. Belloni, Radiat. Phys. Chem. 1989, 34, 605; also Ref. 20b, pp. 198ff.
22. P. Fayet, F. Granzer, G. Hegenbart, E. Moisar, B. Pischel, and L. Wöste, Phys. Rev. Lett. 1985, 55, 3002.
23. (a) A. Hoffman, J. Charkoudian, and A. Ames, Photogr. Sci. Eng. 1974, 18, 12; (b) A. Hoffman, Thermodynamic Theory of Latent Image Formation, Focal Press, London, 1982.
24. (a) J. M. Hedges and J. W. Mitchell, Philos. Mag. 1953, 7, 223; 357; (b) K. Starbova and V. Platikanova, Photogr. Sci. Eng. 1978, 22, 6; 1979, 23, 107.
25. (a) T. H. James, in Theory of the Photographic Process, 3rd ed., T. H. James, Ed., Macmillan, New York, 1966, p. 352; (b) P. J. Hillson, J. Photogr. Sci. 1975, 23, 15.
26. M. R. V. Sahyun, Photogr. Sci. Eng. 1974, 18, 383.
27. H. W. Wood, J. Photogr. Sci., 1966, 14, 72.
28. M. Accola, M. B. Mizen, J. R. Pladziewicz, and M. R. V. Sahyun, unpublished results.
29. E. Weyde, Photogr. Korresp. 1961, 98, 7; G. C. Farnell, and L. R. Solman, J. Photogr. Sci. 1970, 18, 94.

30. S. Fukuzumi, C. L. Wong, and J. K. Kochi, *J. Amer. Chem. Soc*. 1980, *102*, 2928.
31. (a) R. G. Willis, F. E. Ford, and R. B. Pontius, *Photogr. Sci. Eng*. 1970, *14*, 141; (b) R. G. Willis and R. B. Pontius, *ibid*., 149.
32. J. R. Pladziewicz, P. Slattum, and M. R. V. Sahyun, *J. Imaging Sci*. 1989, *33*, 119.
33. R. L. Bent, J. C. Dessloch, F. C. Duennebier, D. W. Fassett, D. B. Glass, T. H. James, D. B. Julian, R. W. Ruby, J. M. Snell, J. H. Sterner, J. R. Thirtle, P. W. Vittum, and A. Weissberger, *J. Amer. Chem. Soc*. 1953, *73*, 3100 (1953); see also Ref. 25a, pp. 294ff.
34. V. I. Levich, in *Physical Chemistry, An Advanced Treatise*, vol. IX, H. Eyring, Ed., Academic Press, New York, 1970; chap. 12; M. R. V. Sahyun, *J. Photogr. Sci*. 1983, *31*, 243.
35. P. Mulvaney, T. Linnert, and A. Henglein, *J. Phys. Chem*. 1991, *95*, 7843.
36. D. D. F. Shiao and T. Bistrovich, *J. Imaging Sci*. 1991, *35*, 279.

Faster and Faster Colour Processing

J. R. Fyson

RESEARCH DIVISION, KODAK LTD., HARROW, MIDDLESEX HA1 4TY, UK

1 INTRODUCTION

This paper reviews some of the chemistry changes that have taken place in colour films and processing chemistry to enable photographs to be produced from colour negatives in shorter times.

There has been a desire since the beginnings of photography for more rapid processes. For people processing their films by hand, there is the relief from much of the boredom of keeping an eye on the time, remembering to turn a tank over once every 30 seconds or so; and for the large commercial photofinisher the ability to get prints for customers more quickly. By shortening the process the efficiency is also improved. For a given geometry of processing machine the following relationship between parameters applies:

$$\text{Process time} = \frac{\text{(Film path length + Film length)}}{\text{Transport Speed}}$$

If the process time is reduced as occurred in the early 1970s, more films can be put through a machine in a given time, by speeding up the drive motor. Alternatively one can run a machine at the same transport speed but make the film path length and therefore the machine, smaller. This machine could be made more cheaply. It was thus possible for more people to start up processing businesses with less capital outlay. At first it was necessary to splice films together in the dark prior to processing. However machine design changes towards the end of the 1970s and beginnings of the 1980s, allowed films to be processed individually in daylight loading machines that we often find in minilabs which we see in many high street stores. These can provide customers with prints within an

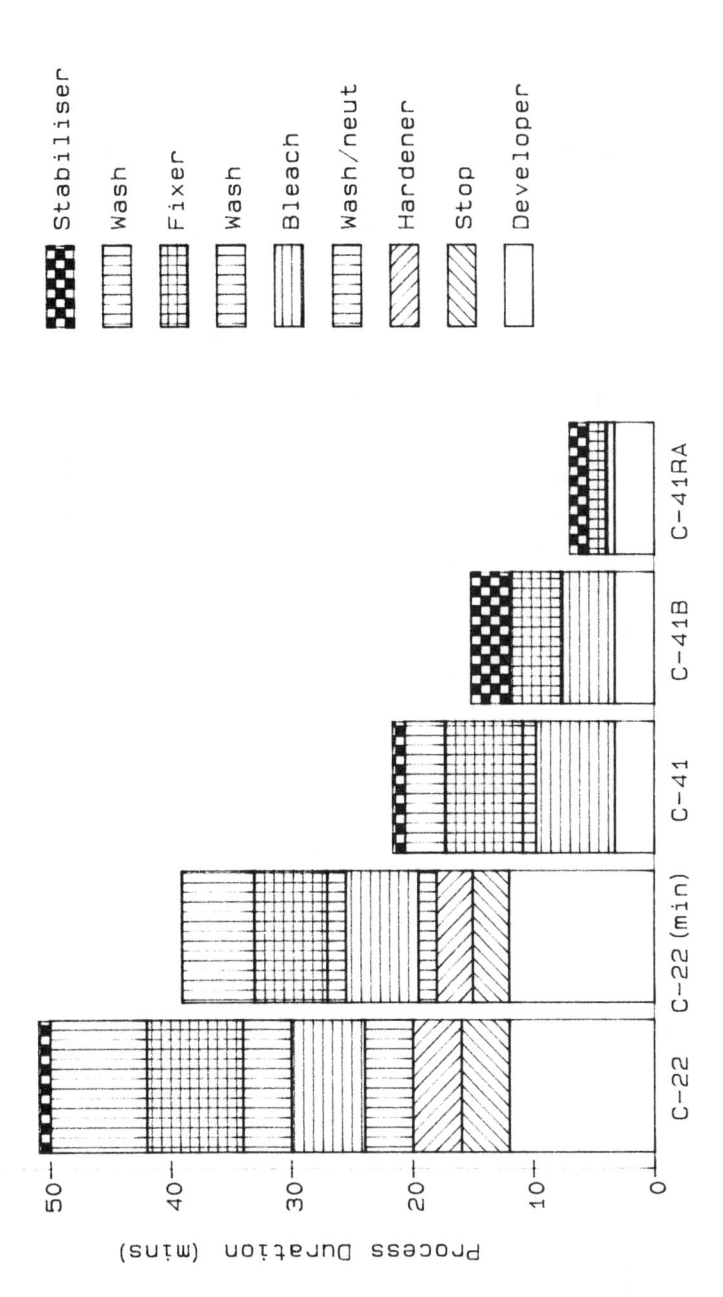

Figure 1 Comparison of Duration of Different Colour Negative Processes

hour of the exposed film being handed in, as they are processed in the shop.

This paper will consider the colour negative process as representative of the changes that have taken place in the chemistry that allows silver halide photographic materials to be processed more rapidly. The following KODAK Color Film Processes will be discussed:

KODAK Color Film Processing Chemicals, Process C-22
KODAK FLEXICOLOR Chemicals (Process C-41)
KODAK FLEXICOLOR RA Chemicals (Process C-41)

Figure 1 shows the relative process times and the steps in the processing of KODACOLOR films since 1949[1,2,3,4]. The C-22 process is similar to the one used for KODACOLOR film since its inception in the early 1940s. The original process took just under one hour and consisted of 9 wet processing steps, the longest being the developer taking 12 minutes or so. The actual time was film and processing machine dependent. This process was in common use until 1975.

2 THE C-22 PROCESS

Developing

The main processes taking place in the developer can be represented by the following general equations. A more detailed account of the reactions is reviewed by Tong elsewhere[5].

$$CD + AgX \longrightarrow Ag + CDox^+ + X^-$$
colour developer, oxidised developer

$$CDox^+ + R_1\text{--}CL^-\text{--}R_2 \longrightarrow \text{leuco dye} \xrightarrow{-HL} \text{dye}$$

The activation energy of these reactions has been measured. The effect of temperature on the first reaction was measured in film by Birki and Jenny[6], who found that a 10°C rise in temperature gave an increase in development rate between 1.9 and 3.7 times, depending on the developer constitution and the film's exposure. This corresponds to an activation energy of between 50 and 106 kJ/mol.

The activation energy of the second reaction has been measured by stopped flow techniques with the coupler dispersed in an oil phase; the form used in most Kodak films. It was found to be between 40 and 100 kJ/mol. It

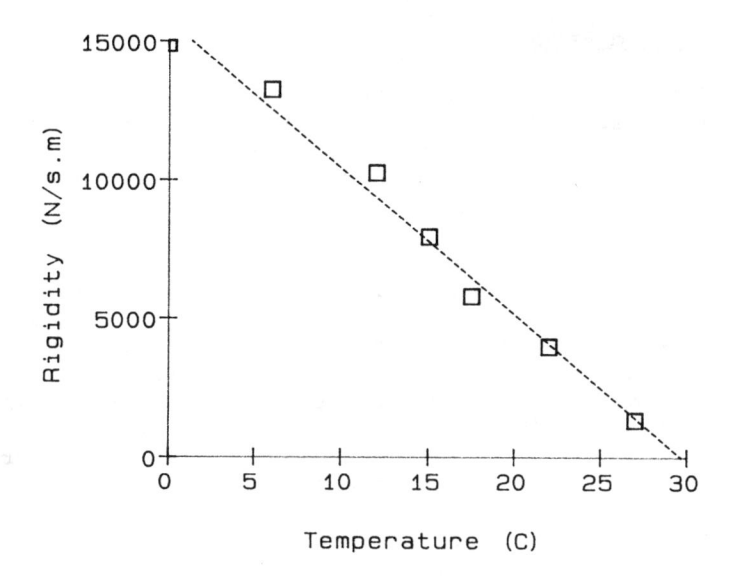

Figure 2 Rigidity of a 57.5g/l Gelatin Solution

is found in practice that the first reaction governs the kinetics of development. To reduce the development time it would seem reasonable to increase the processing temperature as we would expect a halving of the process time for every increase of 10°C or so[7].

The silver halide and colour forming materials in the early KODACOLOR films were suspended in a unhardened gelatin matrix. If the development were carried out at temperatures much higher than 24°C, the C-22 process temperature, the swollen wet gelatin would melt before the film was fully developed. Figure 2 shows the rigidity of set unhardened gelatin in water as the temperature is raised. We can see that the gelatin has melted, i.e. has no rigidity, above 30°C[8].

The processing was carried out at 24°C, the highest temperature that is consistent with fast development rate and sufficient rigidity of the gelatin. This means that the development step time was constrained and 12 minutes was necessary to get the correct contrast for printing.

Stopping and Hardening

The development reaction was closed down by following the developer with an acid stop bath. This reduced the pH in the layers, protonated the coupler, preventing further dye formation, and it diluted the developer components[9].

Even at 24°C the gelatin becomes soft and easily damaged after prolonged contact with the processing solutions. In order to get around this problem the gelatin was hardened by passing the film through an alkaline bath containing an aldehyde, usually formaldehyde[10]. The formaldehyde reacts with the primary amine groups in the side chains in the gelatin, particularly those of lysine and hydroxylysine, to cross-link neighbouring gelatin strands[10].

This has the effect of reducing the amount of water taken up by the gelatin, which is often measured by the swell of the film. It also raises its melting point, and makes it more resistant to abrasion[12].

The hardening bath could not be used before the developer because the formaldehyde reacted with the couplers coated in the film and rendered them inactive in the developer. Particularly sensitive were the magenta pyrazolone couplers[13].

The formaldehyde in the hardening bath following the developer used this reaction to advantage. The formaldehyde reacted with any unreacted magenta couplers in the film. This was found to stabilise the magenta dyes that had formed in the developer[14], as it had been found that unused coupler would react with dyes formed to give a colourless or yellow compound on prolonged storage.

Excess formaldehyde in the hardener was washed out preventing too much cross-linking of the gelatin. The cross-linking was thus controlled such that hardened in this way the gelatin would allow rapid diffusion of chemicals through itself without their passage being severely impeded by the joined collagen strands. The gelatin had to be hard enough that it would remain intact in all the remaining baths of the process and not be scratched.

Bleaching and Fixing

The remainder of the processes, sometimes referred to as the tailends, or complementary processes, the bleach, fix and wash steps, were relatively rapid. The bleach step was required to remove the dark coloured silver formed in the developer and to leave the dye behind. This used a combination of an alkali metal hexacyanoferrate(III) and a bromide to provide a means of oxidising the silver back to silver bromide. This can be simply expressed in the following equation[14]:

$$Ag + Fe(CN)_6^{3-} + Br^- \longrightarrow AgBr + Fe(CN)_6^{4-}$$

A fixer was required to remove the light sensitive silver salts from the film. These were either those silver halides that were not developed or formed by oxidation of developed silver in the bleach[15]. Without removal the silver halide would slowly turn dark in the light and would scatter light in an enlarger making prints difficult to make.

The fixer used sodium thiosulphate as the silver halide solvent. The reaction is of the form

$$AgX + nS_2O_3^{2-} \longrightarrow Ag(S_2O_3)_n^{(2n-1)-} + Br^-$$

where n= 2-4 depending on the conditions and the cations present[16].

The fixer was stabilised by the addition of metabisulphite. This probably works by forcing the equilibrium of the following two-phase reaction to the left[17].

$$S_2O_3^{2-} + H^+ \longleftrightarrow HSO_3^- + S$$

After fixing the film was washed to remove any remaining soluble chemistry in the film which otherwise would cause the dyes to fade or crystals to form in or on the gelatin layers.

Following the wash the film was given a quick rinse in a stabiliser which contains a wetting agent which allowed the film to dry in the hot air of a drier without showing washing streaks caused by hard water.

A fast version of the C-22 process was used where speed was of the essence. This left out the safety time factors built into the process and some of the wash steps which made reuse of the chemicals easier. The cost of doing this was a less stable film and the need to use more chemistry.

A wash was necessary between the bleach and the fixer. If this was omitted bromide would be carried into the fixer and would reduce its efficiency; and the iron would make it difficult to carry out any silver recovery and reduce the fixer stability by oxidising the bisulphite stabiliser.

3 THE C-41 PROCESS

The slow step in the C-22 process was the development, although the large number of shorter processes contribute to the major part of the processing time. The process time could be best reduced by reducing the development time and perhaps reducing the number of processing steps.

Hardening

It had been established that the activation energy of the development reactions is between 40 and 106 kJ/mol. If the temperature of the development process could be increased a valuable amount of time could be salvaged. However raising the process temperature would cause the gelatin to melt. One way around this was to raise the aqueous melting point of the gelatin. This could be done in two ways: the hardening bath could be put before the developer, provided that a hardener could be found that did not react with the couplers in the film. This was done in the KODAK 'Ektachrome' Film Processing Chemicals, Process E4[18]. The hardening bath contained a dialdehyde which cross-linked with the gelatin, making it less swellable and raising its melting point[19]. The hardening obtained in this way was adequate for processing up to 30°C but the hardener tended to form internal cyclic compounds with the collagen without cross-linking.

A different approach was taken in the negative process. A number of compounds were discovered, that could be added to gelatin solutions before they were coated onto the base, and that would cross-link with the gelatin. A number of different types were available including carbonyls, sulphonate esters, triazines, active halogen compounds, epoxides and activated olefins. These are described in numerous references obtained from 'The Theory of the Photographic Process'[20]. These compounds did not interfere with the ingredients in the film and could be mixed with the gelatin some time before coating. The hardening action, which is believed to work by the formation of cross-links between lysine, hydroxylysine and histidine groups of adjacent gelatin molecules, is very slow in dilute solution[21]. After the gelatin is coated, it is chilled and then dried in warm air. As the water is removed, so the concentration of the hardener and active gelatin sites increases. The temperature of the gelatin/hardener mixture is also increased. The reaction rate increases and hardening takes place. The system is engineered such that most of the hardening takes place before the film leaves the drier. If this does not happen and the hardening reaction continues in the film can, film parameters that depend on the hardness and swellability of the gelatin layers change with keeping of the film stock.

'Forehardening' as this hardening method is called has another benefit. It limits the amount that the gelatin can swell in processing solutions, which reduces the distance that processing species diffuse to get to their site of reaction. This should speed up processes that are diffusion dependent, particularly the bleach and fix, by reducing the diffusion path length to the reaction sites. However, if too much hardener is included there is too little swell and the passage of the active species is impeded by the gelatin strands. Green[22] and Fyson[23] have measured the diffusion constants of different active processing species in hardened gelatins: see Figures 3 and 4. From these the optimum gelatin 'swellability' can be determined for developers, bleaches and fixers and they

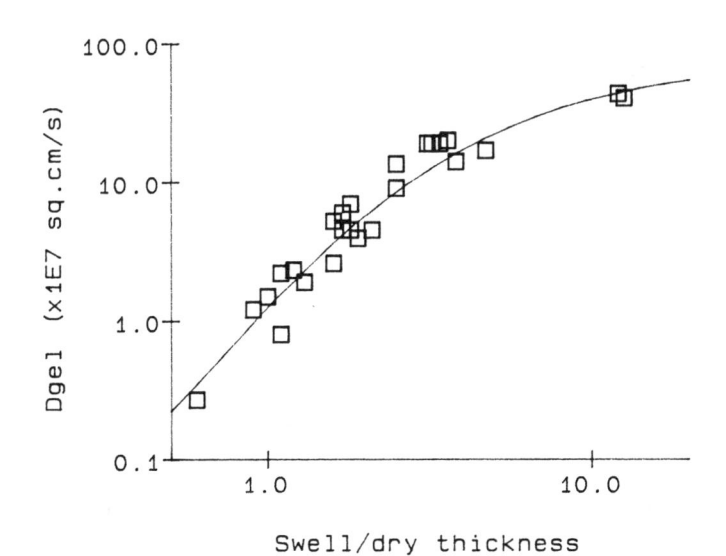

Figure 3 Dependence of Diffusion of Thiosulphate Through Gelatin on its Hardness

Figure 4 Dependence of Diffusion of Bleaching Agents Through Gelatin on its Hardness

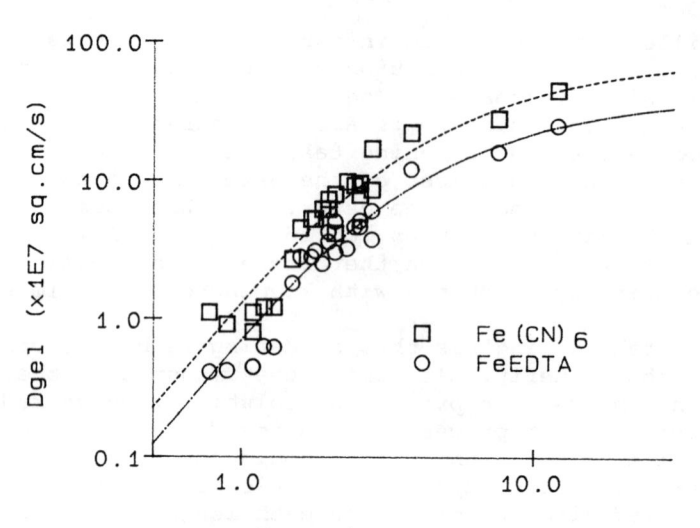

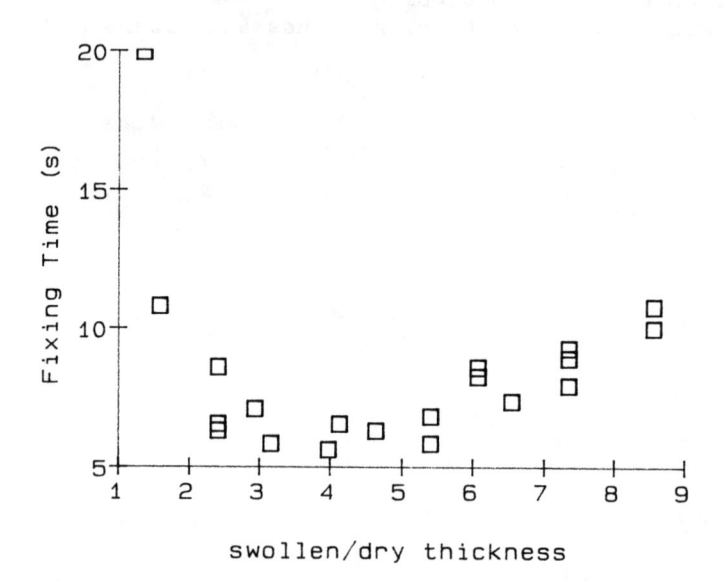

Figure 5 The Relationship Between Gelatin Coating Swellability and Fixing Time

Table 1 Electrode Potentials of Developing Agents

Process	Developing Agent	$E_{1/2}$(pH10)[25] versus NHE
C-22	CD-3	256mV
C-41	CD-4	261mV

have shown that for maximum process speed, the gel should be hardened such that it swells to between 2.5 and 4.5 times its original volume. Figure 5 shows an example of the variation of fixing speed with swellability of a gelatin coating[22]. This is the swellability that the films are usually hardened to.

Developing

As well as forehardening films, allowing the higher developing temperature of 37.8°C of the development step of the C-41 process, other changes were made. The developing agent was changed to a slightly more active one[24], KODAK Color Developing Agent CD-4. This had a similar reducing electrode potential but the oxidised form coupled to give dye more rapidly and efficiently than KODAK Color Developing Agent CD-3.

Bleaching

The bleaching agent was changed from hexacyanoferrate(III) to iron (III) EDTA. This was not such a powerful oxidising agent, see Table 2, but had two advantages. It could be used immediately after the developer without any intervening stop and wash baths as developer carried over in the film did not oxidise. The oxidation by hexacyanoferrate(III) is severe and causes unacceptable amounts of staining.

This new bleach and the elimination of the hardening bath, made unnecessary by the forehardening of the film, removed three steps from the process.

Another advantage was that after bleaching, the iron (II) complex formed could be reoxidised by blowing air through the solution, thus reducing the amount of iron compound that had to be supplied in the replenisher to maintain bleach activity. This was achieved by having some sort of air sparging device in the processing tank.

$$Ag^0 + Fe(EDTA)^- \; Br^- \longrightarrow AgBr + Fe(EDTA)^{2-}$$

$$4Fe(EDTA)^{2-} + 4H^+ + O_2 \longrightarrow 4Fe(EDTA)^- + 2H_2O$$

Table 2 Use pHs and Electrode Potentials of Bleaching Agents

Bleaching Agent	Bleach pH	$E_{(1/2)}$ versus SCE[23]
Hexacyanoferrate(III)	6.0	+170mV
Iron(III) EDTA	6.0	-120mV
Iron(III) 1,3-PDTA	4.75	+5mV

The duration of the less active bleach time was kept short at 6.5 minutes by raising the temperature to 38°C.

Fixing

The fixer was improved by changing from sodium thiosulphate to ammonium thiosulphate. Levenson showed that with typical films fixing was up to 4 times faster using the ammonium salt than the sodium one. Figure 6 shows the data of Kawasaki et al that Levenson[26] used to demonstrate this. The fixing temperature was also raised to 38°C.

Washing

A wash tank was still put between the bleach and the fixer, to prevent iron build up in the fixer. This made it easier to recover the silver from the fixer electrolytically and to reduce the likelihood that the film dyes would be reduced by iron(II) species. In later versions of the C-41 process, C-41B and C-41RA, this step was left out, particularly in sites where electrolytic silver recovery was considered too difficult or uneconomic to be practised, such as in small processing laboratories. The higher wash water temperature (38°C), and the optimised hardening which allowed more efficient removal of unwanted components remaining in the film from the process, allowed the washing time to be reduced.

4 FURTHER IMPROVEMENTS

All these modifications combined to give a C-41 process with a total wet process time of 22 minutes[2]. Although this shortened time allowed the production of a small colour processing machine which could be used by smaller companies to develop film, there was no equivalent improvement in the paper processing or combined printer and paper processor that could be operated in the dark, so there were still no high street minilabs. Small operations started, often running from cellars and garages. The now familiar minilabs came when integral printers and processors were produced. These were first seen in an exhibition in 1979.

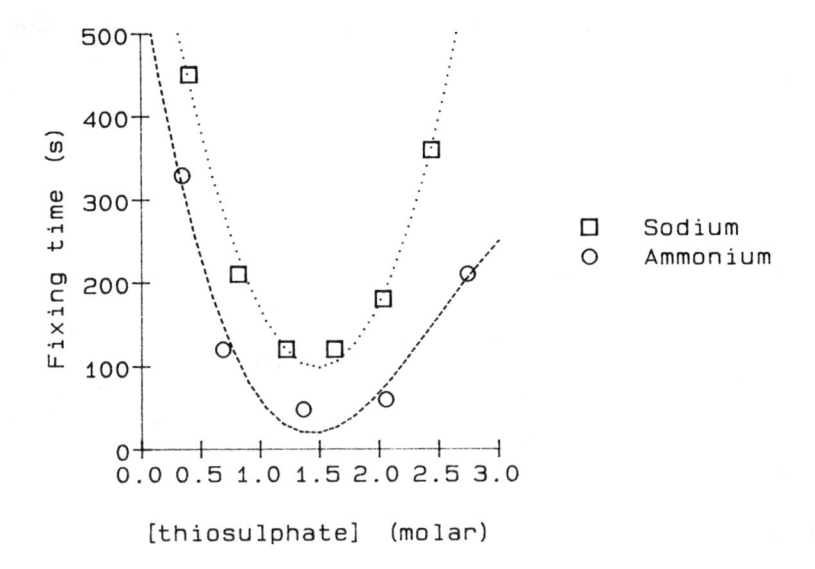

Figure 6 Effect of Thiosulphate Concentration on Fixer Clearing Time

With the advent of minilabs came the need for even more rapid processing. By this time there were many different films that could be processed in C-41 chemistry. Each had been balanced to give the right colour characteristics in the C-41 process, in the 3.25 minutes of the developer time and 37.8°C temperature. The process could often be warmed up and a film given a shorter time that would still give optimum performance for that fim type. However, other film types and makes going through the same process would require a different optimum process time and temperature. Therefore it was impossible to change the developer part of the process to accommodate all films available.

It was necessary to look for other ways of reducing the process time, particularly by making the rest of the process more efficient.

Faster Bleaches

The first change was to reduce the bleach time to 4 minutes. This was found to be adequate for all films in use at the time. The bleach was then changed to one which contained some iron (III) 1,3-propylenetetracetic acid (PDTA) in addition to the iron(III) EDTA[27]. This has a more oxidising electrode potential than the EDTA complex, see Table 2, and is less able to form a slowly diffusing dimer which iron (III) EDTA is prone to do at high concentrations and higher pH[28].

$$2FeEDTA^- + 2OH^- \longrightarrow EDTAFeO_2FeEDTA^{6-} + 2H^+$$

Some EDTA was left in as it appeared to cut down the amount of unwanted dye formed by oxidation of the developer in the bleach. In the RA(Rapid Access) version of the bleach, the last in the series all the EDTA was replaced with PDTA. The stain was prevented by increasing the pH buffering in the bleach by raising the acetic acid concentration. This ensured that any coupler anion in the coating was quickly protonated so that it could not react with any oxidised p-phenylene diamine that might form in the bleach from developer carried over in the film.

Mixed Fixers

There was not much change in the fixer in the C-41 process until the requirement for a very short process, the RA version mentioned above. The original C-41 fixer consisted of a 0.85 molar solution of ammonium thiosulphate stabilised by the addition of some sulphite. If the concentration of the thiosulphate is raised, at first the reaction gets a little faster, but then slows down as the concentration of thiosulphate exceeds about 1.3 molar. If the swell of the film is measured in different thiosulphate concentrations, it is seen to reduce as the concentration goes up, see Figure 7[29] As the gelatin is less swollen at high thiosulphate concentrations the gelatin molecules impede the diffusion of thiosulphate molecules, and this effect outweighs the effect of the increased concentration. This has been studied in detail by Kawasaki et al. Figure 6[31] shows the overall effect of changing the thiosulphate concentration on fixing rate of a film.

One way around this is to find a complexing agent for silver that does not deswell gelatin to the same extent as thiosulphate. One such ion is thiocyanate. This strongly complexes silver and does not deswell gelatin, in fact it has been shown to swell it[30]. However, silver thiocyanate is almost insoluble in water (2.1×10^{-4} g/l at $25°C$)[31], but soluble in thiosulphate solutions[32]. The answer to this fixing problem was to use as the fixer a mixture of thiosulphate and thiocyanate. Getting the proportions right gave a fixer that would remove all the silver halide in all current films in 90 seconds and with some materials in even less time. By reducing the washing to the bare

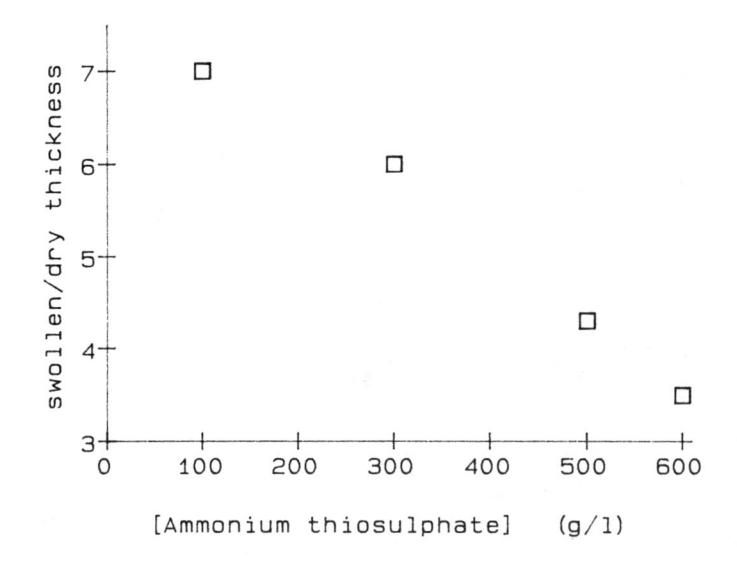

Figure 7 Effect of Thiosulphate Concentration on the Swellability of Coated Gelatin Layers

minimum but still retaining good image stability a process of 7.5 minutes wet time could be produced.

5 SUMMARY

The processing time for colour negative films has been reduced from just less than one hour to 7.5 minutes. This has been achieved by finding ways of raising the melting point of gelatin so that the process temperature can be raised, accelerating all the chemical reaction rates. This also removed the need for a hardening bath. A change of bleaching agent type allowed for the removal of two more baths from the process. Further cuts were made to the process and a change of fixing agent first from sodium thiosulphate to the ammonium salt and then the addition of thiocyanate reduced the fixing time by 84%.

It is difficult to see how the process time might be reduced still further as most of the processes are being governed by diffusion of material in the film. Increasing the temperature has little effect on diffusion rates and the energy requirement for the process may begin to be prohibitive.

REFERENCES

1. KODAK Publication Z-102, "Monitoring System for Processing Kodacolor-X"
2. KODAK Publication Z-121, "Using Process C-41"
3. KODAK Film Processor, System 25 Manual
4. KODAK Film Processor, System 52R Manual
5. L.K.J.Tong; T.H.H.James(ed); "The Theory of the Photographic Process", 4th edition, p 339 (Kodak, 1977)
6. F.Birki, L.Jenny; Helv.Chim.Acta, 1943, 26, 2264
7. J.Willder; Private Communication, 1992
8. J.D.Ferry; J.Am.Chem.Soc., 1948, 76, 2244
9. a) K.H.Stephen; T.H.H.James(ed); "The Theory of the Photographic Process", 4th edition, p 462 (Kodak, 1977)
   b) L.F.A.Mason; "Photographic Processing Chemistry" p 258 (Focal Press 1966)
10. T.H.H.James(ed); "The Theory of the Photographic Process", 4th edition, p 79 (Kodak, 1977)
11. a) K.H.Gustavson; "The Chemistry of Tanning Processes" (Academic Press, New York, 1956)
    b) H.Fraenkel Consat, A.S.Olcott; J.Am.Chem.Soc., 1946, 68, 34
    c) Ibid. 1948, 70, 2673
12. R.W.G.Hunt; "The Reproduction of Colour", 2nd edition, p 306 (Wiley and Sons Inc., New York, 1967)
13. P.W.Vittum, F.C.Duennebier; J.Am.Chem.Soc., 1950, 72, 1536
14. G.I.P.Levenson, T.H.H.James(ed); "The Theory of the Photographic Process", 4th edition, p 448 (Kodak, 1977)
15. G.I.P.Levenson, Ibid. p 437
16. J.Pouradier, A.Pailliotet, C.R.Barry; Ibid. p 10
17. G.I.P.Levenson, Ibid. p 438
18. KODAK Publication Z-100, "Monitoring System for Processing Ektachrome Films", Process E4'
19. a) J.H.Bowes, C.W.Cater; Biochim.Biophys.Acta 1965, 168, 341
    b) J.H.Bowes, C.W.Cater; J.App.Chem., 1965, 15, 29619.
    c) J.H.Bowes, C.W.Cater; J.Am.Leather Chem.Assoc. 1965, 60, 275
20. D.M.Burness, J.Pouradier, T.H.H.James(ed); "The Theory of the Photographic Process", 4th edition, p 81-84 (Kodak, 1977)
21. a) V.L.Zelikman; Zh.Nauchn.Prikl.Fotogr.Kinometogr. 1960, 5, 403
    b) Ibid. 1961, 6, 63
    c) Ibid. 1964, 9, 205
22. A.Green, G.I.P.Levenson; J.Photogr.Sci., 1976, 24, 193
23. J.R.Fyson; J.Photogr.Sci., 1984, 32, 234
24. "British Journal of Photography Annual 1976", p 213

25. a) L.K.J.Tong, M.C.Glesmann; Photogr.Sci.Eng., 1964, 8, 319
    b) R.L.Bent et al; J.Am.Chem.Soc., 1952, 73, 3100
26. M.K.Kawasaki, T.Ohtsu; J.Soc.Phot.Sci.Tech.Japan, 1972, 35, 361
27. C.M.Macdonald, K.H.Stephen; Research Disclosure 24033
28. a) R.L.Gustavson, A.E.Martell; J.Phys.Chem., 1963, 67, 567
    b) H.J.Shugar, A.T.Hubbard, F.C.Anson, H.B.Gray; J.Am.Chem.Soc., 1969, 90, 71
29. D.B.Alnutt; J.Soc.Motion.Pict.Eng., 1943, 41, 300
30. S.E.Sheppard, F.A.Elliot, S.S.Sweet, J.Franklin; J.Franklin Inst., 1923, 196, 445
31. "Handbook of Chemistry and Physics", 70th edition (CRC Press, 1989)
32. a) R.C.M.Smith, E.R.Townley; J.Photogr.Sci., 1959, 7, 55
    b) E.T.S.Smith; British Patent 931,294

Noble Metals for Common Images

Michael J. Ware

DEPARTMENT OF CHEMISTRY, UNIVERSITY OF MANCHESTER, MANCHESTER M13 9PL, UK

ABSTRACT. The photochemistry of iron(III) polycarboxylates offers a route into the photographic printing of images in noble metals, whereby the iron(II) photoproduct can reduce complexes of platinum, palladium, gold and silver to an image constituted of the metal in the colloidal state. The history of these processes is briefly reviewed, and up-dated methods are described for the provision of contact prints in these metals to serve specialist photographic practices, especially in the areas where archival permanence is paramount. Chemical rationalizations are offered to explain the characteristics of these 'alternative' printing processes which distinguish them from the usual gelatin-silver halide medium.

1   INTRODUCTION

In the entire field of photochemistry, the sensitivity of silver halides with development seems uniquely high and provides, as yet, the only practical 'camera-speed' materials for capturing the photographic negative. Printing the positive, however, does not necessarily demand high photosensitivity and can be achieved by diverse processes, many of which have been known[1] since the earliest days of photography, but have either fallen into disuse or never been carried into successful practice in the first place. Commercial photographic printing today is naturally dominated by the gelatin-silver halide papers manufactured by the major industry to previously unparallelled standards of convenience, speed and technical quality. But the choice of commercial print-making materials was once much wider than it is now: at the turn of the century, for instance, more prints in platinum than silver were to be seen[2] on the salon walls. Although the 'alternative' photographic printing processes cannot compete today with the silver-gelatin 'monoculture' they should not therefore be regarded as totally obsolete, because they can offer unusual qualities of interest to certain specialist photographic practices.

Historical Background

In the beginnings of photography it was by no means obvious which photochemical system was to provide the wellsprings for a mainstream of successful technical development; as Herschel remarked[3] in 1839 "...*I was on the point of abandoning the use of silver in the enquiry altogether and having recourse to Gold or Platina...*" Many photosensitive materials were consequently examined, both organic and inorganic, including silver, gold and platinum salts, anthocyanine dyes, dichromated colloids and iron(III) carboxylates; it is this last category that is reviewed here.

Döbereiner in 1831 was the first to observe[4] the light-induced decomposition of iron(III) oxalate. Photosensitivity has since been discovered in other carboxylates of iron(III), notably the citrate, malonate, tartrate and glycollate complexes; although no clear criterion has yet emerged for deciding what structural feature of such complexes is necessary for photosensitivity. On irradiating the complex with light having some blue or ultra-violet content, electron transfer occurs: the iron(III) is reduced to iron(II) and the ligand is oxidised, usually with decarboxylation. The iron(II) photoproduct, which is very reactive, can then be made to yield a permanent image in three ways:
(i) Reduction of a noble metal complex salt to the finely divided metal, *e.g.* Pt, Pd, Au, Ag - giving rise, respectively, to the *Platinotype*, *Palladiotype*, *Chrysotype* and *Argentotype* processes.
(ii) Coupling with $[Fe(CN)_6]^{3-}$ to yield Prussian Blue - the *Cyanotype* or *Blueprint* process, formerly of wide commercial use.
(iii) Leaving iron(III) to form 'inks' with gallic or tannic acid - the *Colas* or *Ferrogallate* process.

This review examines the first of these subcategories: the formation of images in noble metals.

Gold Printing. In 1842 Sir John Herschel described[5] the use of ammonium iron(III) citrate and sodium tetrachloroaurate(III) to obtain purple images in colloidal gold; these may still be seen in the collections of the Museum of the History of Science, Oxford, and the Royal Society Library, London. This *Chrysotype* process was not taken up, however, because of the expense, chemical difficulties, and the competition from processes using silver halides, that were rapidly improving at the hands of William Henry Fox Talbot.

Platinum Printing. Although platinum salts were investigated[3] photochemically as early as 1830 to 1845 with a view to their imaging potentiality, it was not until 1878 that a viable printing process was developed by William Willis,[6] using iron(III) oxalate and potassium tetrachloroplatinate(II), which overcame the problem of the slow reaction rate of platinum redox chemistry. Willis founded the *Platinotype Company* in 1879 in order to market his platinum printing papers commercially[7]. The Company survived until 1937, producing a palladium paper in 1916 when platinum became a strategic material. Platinum-palladium paper was not produced again commercially until 1988, with the formation of the Palladio Company in the USA, which reflected the recent revitalised interest in alternative printing.

2 CHARACTERISTICS OF IRON-BASED PROCESSES

Photochemistry of the *tri*soxalatoferrate(III) anion

Of the several photosensitive carboxylato- complexes of Fe(III), the trisoxalatoferrate(III) anion has been extensively investigated[8], and forms the basis for the well-known chemical actinometer[9]. The quantum efficiency of the process

$$[Fe(C_2O_4)_3]^{3-} \rightarrow [Fe(C_2O_4)_2]^{2-} + CO_2$$

in aqueous solution is approximately unity[10] for radiation of wavelength between 250 and 400 nm, falling sharply thereafter. At 365 nm a typical sensitized layer can absorb 90% of the radiation falling on it. The iron(II) oxalato-complex photoproduct is quite a strong reducing agent ($E^\theta([Fe(C_2O_4)_3]^{3-}/[Fe(C_2O_4)_2]^{2-}) = +0.02$ V) and liberates 'noble' metals from their complexes, kinetic factors permitting, thus:

$$[MX_4]^{2-} + [Fe(C_2O_4)_2]^{2-} \rightarrow M\downarrow + Fe(III)$$

The reducible noble metals are evident from their redox potentials:

$E^\theta([PtCl_4]^{2-}/Pt) = +0.73$ V
$E^\theta([PdCl_4]^{2-}/Pd) = +0.62$ V
$E^\theta(Ag^+/Ag) = +0.80$ V
$E^\theta(Hg^{2+}/Hg) = +0.85$ V
$E^\theta([AuCl_4]^-/Au) = +1.00$ V

Photographic images have been printed in Pt, Pd, Ag, Hg and Au. In principle, Ru, Rh, Re, Os, and Ir are also reducible, but their use for imaging is inhibited by kinetic factors in some cases, and by sheer expense in others.

The photochemical yield of image-forming metal may be calculated[11] for typical sensitizer parameters: it predicts exposure times in the order of two minutes to a UVA light source delivering a flux of 50 W/m² at 365 nm. This prediction is born out in practice.

Optical Properties of Colloidal Metals

The noble metal precipitated by these photochemical means often has a particle size in the nanometer region, *i.e.* it has colloidal dimensions, in contrast to the micron-sized bundles of filamentary metallic silver in most developed silver-gelatin materials. The chemistry and physics of small metallic particles is attracting great interest at the present time[12], although the recognition of metals in the colloidal state causing brilliant colours goes back to Michael Faraday[13] in 1857. With particle sizes less than the wavelengths of visible light, metal sols display interesting optical characteristics due to the excitation by light of collective oscillations in their conduction electrons, which are known in quantum-mechanical language as plasmons[14]. For most metals, such surface plasmon resonance generally gives rise to absorption maxima in the ultra-violet region of the spectrum but in a few cases, notably Cu, Ag and Au, the variation of the dielectric function of the metal with frequency causes quite sharp absorption bands in the visible region, giving rise to the striking colours that first attracted Faraday's attention. A theoretical treatment of this phenomenon for spherical metal particles was given by Gustav Mie[15] in 1908, using Maxwell's electromagnetic theory, and was subsequently extended to ellipsoidal particles by Gans[16]. When the particles are much smaller than the wavelength of light a dipolar approximation is valid and absorption predominates but, at larger radii, multipolar terms become important and light scattering also becomes significant. This topic has been taken up again and greatly extended by Milton Kerker[17]. Most recently, Creighton and Eadon[18] have calculated absorption spectra for 10 nm diameter particles of most of the metallic elements; their work shows that distinctive colour is a relatively uncommon property among colloidal metals: the surface plasmon absorption band only peaks in the visible region for the alkali metals, alkaline earths, coinage metals, and scandium, yttrium, europium and ytterbium. In view of the requirement for a pigment to be chemically inert, it follows that copper, silver and gold are likely to remain the only metals providing distinctive colours for decorative or image-making purposes, other metals being grey or brown in the colloidal state.

    The colours of colloidal metals may be modified by several factors: besides the particle size and departure from sphericity mentioned above, the linear aggregation of spherical particles causes the appearance of a long wavelength absorption band due to the splitting of the degenerate surface dipolar plasmon mode into lateral and longitudinal components, similar to a prolate spheroid. The plasmon absorption band is also shifted by the presence of molecules or ions adsorbed onto the surface of the particle[19], where the metal atoms are coordinatively unsaturated. Henglein[20] has also shown that colloidal silver particles can act as an electron pool towards redox active species, and that the stored charge influences the plasmon resonance absorption. These factors are responsible for some striking changes in the colours of colloidal metal images during their wet-processing procedures.

    The extremely small dimension of the colloidal metal

particles in such images, compared with those in conventional silver photographic materials, suggests that they may have potential for optical information storage, but for this purpose it will be necessary to employ a more homogeneous substrate than paper - which is the next characteristic to be discussed.

The Structure of Cellulose Paper[21]

The purest papers are made from cotton linters, and are constituted of >98% α-cellulose, which is a linear polymer of β-D(+)-glucopyranose, 1,4-linked, with 1500 to 10000 monomer units per chain. The polysaccharide chains hydrogen-bond laterally, with the inclusion of some water molecules, so the cellulose structure[22] can display both crystalline and amorphous regions. Bundles of these chains constitute microfibrils, the basic building units of the tubular fibres. The cellulose fibres themselves have a complex morphology at the microscopic level, providing an extremely heterogeneous (but aesthetically pleasing!) material. Performing photochemistry in paper can be further complicated by the presence of various manufacturers' additives: sizing agents, fillers, buffers, retention and wet strength agents, dyes and optical brightening agents. It is not sufficient, as in paper chromatography, just to employ a completely pure cellulose paper; the presence of a sizing agent, for example, is essential in order to localise the sensitizer and subsequent image in the surface fibres. The paper base cannot simply be regarded as an inert substrate, as it is in commercial photographic papers. Rather, the paper must be viewed as a potentially reactive host matrix for the sensitizer ions, which may chemisorb onto the cellulose. With three hydroxyl functions per monomer unit, there is ample scope for hydrogen-bonding, and carboxylates are known to engage in extensive hydrogen-bonding in the crystalline state.

The water content of cellulose at equilibrium with atmospheres of normal ambient relative humidity[23] is also a vital factor in determining the chemistry, because the molar ratio of water to sensitizer ions in a coated paper at ca. 70% R.H. is about ten to one. This confers sufficient mobility on the iron(II) photoproduct that it can migrate short distances to interact with and reduce the noble metal complex, even in the nominally 'dry' state. Thus the final image is 'printed out' during the exposure, before the 'wet' processing procedure designed to remove excess chemicals. In contrast, if the sensitized paper is thoroughly desiccated there is no 'print-out' and the image is only obtained upon wet development.

This behaviour inclines one towards a view of the sensitized layer in cellulose paper as a 'quasi-solution' state of partially chemisorbed ions having a limited mobility in a local aqueous environment; in marked contrast to the model appropriate for silver halide 'emulsions' where the photosensitive material in the microcrystalline solid phase is suspended in a binder gel.

Our understanding of this 'quasi-solution' state is probably now only at the same stage as the understanding of silver-gelatin emulsions was fifty years ago, because relatively little work has yet been done on the study of inorganic ions adsorbed on cellulose.

Summary of Characteristics

The physico-chemical differences between the iron-based processes and modern silver-gelatin printing have practical consequences that may be summarised as follows:
1) Aqueous sensitizers are applied to plain paper.
2) No colloidal binder layer is present: a totally matte surface results. The image substance lies within the surface fibres of the paper.
3) There is no light 'amplification' by development of a latent image; 'speed' is so low that projection printing by an enlarger is not possible with normal technology, only contact printing, and a UVA light source is required. The requirement for a large negative is disadvantageous, but in compensation there is no need for a photographic darkroom.
4) The image can be obtained in a range of noble metals as colloidal dispersions, *viz.* Pt, Pd, Ag, Hg, Au, generating different image colours. The optical resolution is in principle very high.
5) 'Printing-out' of the image is possible, which makes the estimation of correct exposure very simple.
6) There is a wide range of choice in paper surface texture and base colour.
7) Archivally permanent images are achievable.

Such characteristics will not commend themselves for the majority of photographic printing, but they can be advantageous in the areas of archives and fine art.

3 THE PLATINUM & PALLADIUM PRINTING PROCESSES

The traditional platinotype process invented by Willis[6] used iron(III) oxalate as the photosensitive component; on irradiation this yields the highly insoluble iron(II) oxalate which cannot reduce $[PtCl_4]^{2-}$ to Pt metal unless excess ligand, *e.g.* $C_2O_4^{2-}$, is applied in aqueous solution to solubilise the iron(II) oxalate by formation of the complex anion:

$$Fe_2(C_2O_4)_3 \xrightarrow{h\nu} Fe(C_2O_4)\downarrow + CO_2\uparrow$$
$$Fe(C_2O_4) + [PtCl_4]^{2-} + C_2O_4^{2-} \rightarrow [Fe(C_2O_4)_3]^{3-} + Pt\downarrow$$

The coated paper was thoroughly desiccated and, on exposure, gave only a pale image in iron(II) oxalate; this had to be developed in a bath of hot (76°C) potassium oxalate solution. Development at lower temperatures tended to give inferior results owing to the slow reduction kinetics, unless steps were taken to inhibit the rapid dissolution of imaging substance from the paper.

An Updated Method

By contrast, a more convenient method[11] can be based on ammonium trisoxalatoferrate(III) which under conditions of normal relative humidity provides a 'print-out' image because the immediate photoproduct is a soluble complex of iron(II) which can migrate sufficiently to react immediately:

$$[Fe(C_2O_4)_3]^{3-} \xrightarrow{h\nu} [Fe(C_2O_4)_2]^{2-} + CO_2 \uparrow$$

$$[Fe(C_2O_4)_2]^{2-} + [PtCl_4]^{2-} \xrightarrow{H_2O} [Fe(C_2O_4)_2]^{-} + Pt \downarrow$$

No developer is needed, just a clearing agent, disodium EDTA, to remove excess iron(III). In choosing a paper substrate for this method of platinum printing it is very important to avoid those that are sized with gelatin: the ability of proteins to bind strongly to platinum(II) is well-known[24], and the resulting complex is not readily reducible by the iron(II) photoproduct. Paper sized with polysaccharides or alkyl ketene dimer (*Aquapel*) seems compatible with the process.

Mechanistic Aspects

Distinct differences have been noted[11] between the characteristics of platinum and palladium printing: in short, palladium is easier to print and achieves a finer quality under a wider range of conditions. This is due, in part, to the more labile solution chemistry of palladium providing faster reduction reactions, but further observations have led to the hypothesis that the state of aquation of the complex anion is also of critical importance in facilitating the path of the redox reaction. Formation of an aquatrichloro-anion is given by:

$$[MCl_4]^{2-} + H_2O = [MCl_3(H_2O)]^{-} + Cl^{-}$$

The equilibrium constant defined by:

$$K_{aq} = [MCl_3(H_2O)^{-}][Cl^{-}]/[MCl_4^{2-}]$$

has the values[25] at 20°C: $K_{aq} = 0.17$ for M = Pd and 0.015 for M = Pt; thus at the concentration (0.7 M) used for a sensitizer, >50% of the Pd is aquated, but <20% of the Pt is aquated. Scavengers for chloride ions can promote the aquation of Pt(II), which explains the success of Hg(II) additives for improving platinotype formulae:

$$[PtCl_4]^{2-} + Hg^{2+} + H_2O \rightarrow [PtCl_3(H_2O)]^{-} + HgCl^{+}$$

Lead(II) salts may behave similarly, and both were much used by Willis.

The mechanism of the electron transfer reaction between iron(II) and platinum(II) has not been elucidated, but in view of the labile nature of iron complexes to ligand substitution it is unlikely to proceed via an 'inner sphere' mechanism, whereby the two coordination shells share a common bridging ligand. It is more probable that an 'outer sphere' mechanism operates, but this may still find a facile path via the partially aquated complexes[26]: if these are linked by a symmetrical hydrogen bridge, then hydrogen atom transfer from $[Fe(C_2O_4)_2(H_2O)_2]^{2-}$ to $[(H_2O)PtCl_3]^-$, counterbalanced by proton transfer in the reverse direction, results in a net electron transfer to form a transient Pt(I) species:

$$(C_2O_4)_2Fe^{2-} \cdots H \quad H \cdots O \cdots O \cdots PtCl_3^- \quad \rightarrow \quad (C_2O_4)_2Fe^- \cdots H \quad H \cdots O \cdots O \cdots PtCl_3^{2-}$$

It is even possible that water molecules may mediate the process; a further transfer involving a second iron(II) complex is required to complete the reduction.

4 THE GOLD PRINTING PROCESS

Gold in its colloidal state has been used for decorative purposes since the mid 17th century, as the *Purple of Cassius*[27], in red glass and ceramic stains. In the beginnings of photography, Herschel's gold printing process - *Chrysotype*- never came to fruition, because it was eclipsed by less costly, and more tractable processes employing silver. However, gold has always found a place in photography as a print toning agent[28] and a sensitizer for emulsions[29].

<u>Previous Obstacles</u>

The chief difficulty in applying the iron-based imaging system to gold printing arises from the strongly oxidising nature of the most commonly available gold complex, tetrachloroaurate(III), for which:

$$E°([AuCl_4]^-/Au, 4Cl^-) = +1.00 \text{ V}$$

This redox potential is high enough to ensure oxidation of the carboxylate anion, e.g. citrate or oxalate, which is present in the mixed sensitizer, with consequent premature deposition of gold metal. The alternative procedure of placing the gold(III) salt in the development bath, as adopted by Herschel, is now too costly to contemplate.

An additional disadvantage of gold(III) arises from the stoicheiometry of the reduction

$$3Fe(II) + Au(III) = 3Fe(III) + Au(0)$$

which lowers the quantum yield in the final image by a factor of three, leading to low optical densities. Clearly the solution lies in finding a suitable complex, preferably of Gold(I), which has a diminished redox potential with respect to the metal. Until recent years, the only water-soluble, stable Gold(I) complexes known were, typically, the sulphito, cyano, and thiosulphato species, [Au(SO$_3$)$_2$]$^{3-}$, [Au(CN)$_2$]$^-$, and [Au(S$_2$O$_3$)$_2$]$^{3-}$. But these have formation constants so high that the redox potentials are lowered to values less than that of the iron photoproduct, e.g.

$$E^o([Au(CN)_2]^-/Au,2CN^-) = -0.6 \text{ V}$$
$$E^o([Fe(C_2O_4)_3]^{3-}/[Fe(C_2O_4)_2]^{2-}) = +0.02 \text{ V}$$

so no reduction to the metal can occur.

A Suitable Gold(I) Complex

A Gold(I) complex of intermediate stability is required, which is highly soluble in water and reducible by iron(II), but non-reactive with the iron(III) complex in the sensitizer. A ligand satisfying these criteria was found to be 3,3'-thiodipropionic acid, S(CH$_2$CH$_2$COOH)$_2$, used as its soluble sodium or ammonium salts. The reaction of this ligand with gold(III) follows the mechanism already well established[30] for other thioethers, and takes place in three stages[31]:
(1) Coordination with the gold(III).
(2) Reduction of the gold(III) to gold(I), the ligand being itself oxidised in the presence of water to the corresponding sulphoxide.
(3) Coordination of excess ligand to stabilise the gold(I) in aqueous solution against disproportionation.

The following equations illustrate these steps in a schematic way only; the precise species present at equilibrium and their states of ionization, will depend on the cation, pH, and metal/ligand ratio:

(1) [AuCl$_4$]$^-$ + S(CH$_2$CH$_2$COOH)$_2$ → AuCl$_3$S(CH$_2$CH$_2$COOH)$_2$ + Cl$^-$

(2) AuCl$_3$S(CH$_2$CH$_2$COOH)$_2$ + H$_2$O → [AuCl$_2$]$^-$ + Cl$^-$ + 2H$^+$ + OS(CH$_2$CH$_2$COOH)$_2$

(3) [AuCl$_2$]$^-$ + S(CH$_2$CH$_2$COOH)$_2$ → AuClS(CH$_2$CH$_2$COOH)$_2$ + Cl$^-$

The ^{13}C nmr spectrum of the aqueous solution displays only three chemical shifts for the non-equivalent carbon atoms of thiodipropionate, even in the presence of excess ligand, indicating that the ligand exchange rate is rapid at room temperature. Formation of a disubstituted complex in labile equilibrium, is also likely:

AuClS(CH$_2$CH$_2$COOH)$_2$ + S(CH$_2$CH$_2$COOH)$_2$ →
[Au(S(CH$_2$CH$_2$COOH)$_2$)$_2$]$^+$ + Cl$^-$

A New Gold Sensitizer

Hazards. In exploring this chemistry, it is vital to avoid two possible hazards: the formation of the dangerously sensitive explosive, *fulminating gold*, which is precipitated when ammonia is added to gold(III) complexes, and the possible formation of *mustard gas*, $S(CH_2CH_2Cl)_2$, by inadvertant chlorination of the ligand.

Practical Details. For a typical formulation the sensitizer is mixed in the molar ratios:

Ligand : Gold : Iron = from 2 to 3 : 1 : 1

in order to maintain appropriate stoichiometry. Excess ligand provides a more stable but slower sensitizer. Either sodium or ammonium salts can be used. The final pH can be adjusted with the appropriate alkali, but should be kept below pH 7 to avoid undue hydrolysis of the iron(III) complex. The ligand carboxylate functions have a $pK_a \simeq 4$, so tend to provide a source of internal buffer action for the sensitizer.

After exposure, the print is wet-processed in baths at about pH 9, consisting of 5% tetrasodium EDTA and sodium metabisulphite, and finally washed in order to remove all excess chemicals from the paper.

Nature of the Colloidal Gold Image

The colloidal gold so formed is a negatively charged hydrophobic sol which may be protected by the surface adsorption of macromolecules, typically hydrophilic colloids such as gelatin, which inhibit flocculation. The state of dispersion of the sol determines its colour, as

Table 1 Colour and Particle Size of Gold Sols

Shape	Size/nm	Colour
Spherical	<3	Pale blue
	12	Pink
	16	Orange
	20-40	Red
	70	Dark magenta
	100-150	Violet
Irregular	200	Light Blue
Ellipsoids	60x90	Purple
Aggregated		Blue

indicated in Table 1, which is condensed from the observations of Frens[32] and Turkevich *et al*[33]. The particle size and state of aggregation is in turn governed by the pH, humidity, and sizing agent in the sensitized paper, and the wet-processing chemistry. The colour of the image may thus be controlled over a considerable range, including a good magenta and a passable cyan; ironically, a yellow gold sol has not been obtained, so full three-colour printing in pure colloidal gold does not yet seem achievable.

5 IRON-BASED SILVER PRINTING PROCESSES

During the period 1842-98 many formulations were suggested[1] for silver printing employing iron(III) oxalate or citrate, but all suffered from the same basic flaw, that they employed the only soluble silver salt commonly available, namely, $AgNO_3$. Such an oxidising anion has the undesirable ability to re-dissolve the silver image, especially when it consists of colloidal particles with a high surface/volume ratio and when the conditions are acidic, as may be seen from the following standard redox potentials:

$E^\theta (NO_3^-, 3H^+/HNO_2, H_2O)$ = +0.94 V
$E^\theta (NO_3^-, 4H^+/NO, 2H_2O)$ = +0.96 V
$E^\theta (Ag^+/Ag)$ = +0.80 V
$E^\theta (Fe^{3+}/Fe^{2+})$ = +0.77 V

Raising the pH can lower the $E^\theta (NO_3^-)$ values to the point where silver is no longer oxidised, so alkaline buffered developing baths were commonly recommended, but their high pH had the adverse effect of hydrolysing the iron(III) complex, with consequent deposition of irremovable iron(III) hydroxide in the paper, the presence of which, in the long term, caused serious deterioration of the silver image. Such processes, variously named *Kallitype*, *Van Dyke*, *Argentotype* and *Sepiaprint*, consequently acquired a poor reputation for image stability and permanence. Care is also needed in the formulation of the clearing baths: ligands which tend to lower $E^\theta (Ag^+/Ag)$ may cause dissolution of the silver image by the oxidising action of iron(III). The susceptibility of colloidal silver to oxidation may well be worse than the Standard Electrode Potential, which applies to the bulk metal, would indicate: Henglein[19] has pointed out that, for 'micro-electrodes' of small metal particles, there is a dependence of E^θ on the aggregation number (or radius of curvature). This will be only slightly lowered, however, for a silver particle of 10 nm diameter, which still contains about 30,000 Ag atoms.

Solution to the Instability Problem

Images in colloidal silver are not intrinsically unstable, provided they are protected from oxidising and acidic impurities. What is needed in the sensitizer is a soluble salt of silver with a non-oxidising anion: silver sulphamate[34], $AgNH_2SO_3$, fulfills this requirement, and is readily prepared by dissolving the oxide or carbonate in sulphamic acid. A sensitizer incorporating ammonium iron(III) citrate and silver sulphamate has a good stability and provides a yellow image on print-out, which transforms into a rich red-brown colour on processing in water and dilute sodium thiosulphate solution[35]. By increasing the humidity of the paper before exposure, the colloidal silver image is obtained in a purplish-grey form, due presumably to a larger particle size: the colours of silver sols in transmitted and scattered light have been calculated by Wiegel using Mie theory[36].

6 COMPARATIVE SUMMARY OF NOBLE METAL IMAGING

All the processes outlined above have a similar *modus operandi* and a characteristic printing exposure range $\Delta \log H$ of *ca.* 2, in contrast to the much shorter ranges of modern silver-gelatin materials. Negatives therefore need to be made with an appropriate optical density range.

The Photometric Equivalent, P, of an imaging substance, and its Covering Power, C, are defined by[37] $C = 1/P = D/M$ where D is the optical density (under defined conditions of illumination and measurement) and M is the mass of metal/unit area. Table 2 shows the relative costs and covering powers of the noble metals described above, and some properties of the images. Regarding inertness, platinum and gold hold the first place on electrochemical grounds, but it should be remembered that platinum black has extremely high catalytic activity, and can bring about a most undesirable increase in the acidity of the paper substrate in polluted atmospheres by catalysing reactions such as:

$$SO_2 + O_2 \rightarrow SO_3$$

Although the platinum image does not suffer, the paper can become seriously embrittled by the internal build-up of sulphuric acid. Colloidal gold, in contrast, is a poor catalyst and as an imaging substance may not suffer from the only disadvantage which mars the excellence of platinum. Gold, with its higher covering power, is also less expensive in use than platinum and can offer a range of image colours.

Table 2 Comparison of Noble Metals for Imaging

Metal	Cost /£g^{-1}	Covering Power /m^{2}g^{-1}	Colour
Ag	1.2	1	black (micron size)
Ag	1.2	4	yellow, red, brown, purplish
Pd	11.3	2	brown, black
Pt	17.3	1	black
Au	24.3	6	red, orange, pink, green, blue, violet

ACKNOWLEDGEMENTS

The author acknowledges the award of a Photographic Bursary in 1984 by Kodak Ltd., which stimulated this work at the outset. Croda Colloids Ltd., are thanked for information and samples of gelatin.

REFERENCES

1. W. Crawford, 'The Keepers of Light', Morgan and Morgan, New York, 1979; J. Kosar, 'Light Sensitive Systems: Chemistry and Application of Non Silver Halide Photographic Processes', John Wiley and Sons, New York, 1965.
2. H.V. Hyde, Photogr. News, 1901, 45, 680.
3. L. Schaaf, Hist. of Photogr., 1979, 3, 47.
4. J.F. Döbereiner, Pharm. Centralbl. 1831, 2, 383.
5. J.F.W. Herschel, Phil. Trans. Roy. Soc., 1842, 181.
6. W. Willis, British Patents, 1873, No. 2011; 1878, No. 2800; 1880, No. 1117; 1887, No. 1681; 1887, No. 16003.
7. L. Nadeau, 'History and Practice of Platinum Printing', Atelier Luis Nadeau, New Brunswick, 1984.
8. G.D. Cooper and B.A. DeGraff, J. Phys. Chem., 1971, 75, 2897, and references cited therein; H. Sato and T. Tominaga, Bull. Chem. Soc. Jap., 1979, 52, 1402.
9. C.A. Parker, Proc. Roy. Soc. (London), 1953, A220, 104; C.G. Hatchard and C.A. Parker, Proc. Roy. Soc., 1956, A235, 518.
10. J.H. Baxendale and N.K. Bridge, J. Phys. Chem., 1955, 59, 783.
11. M.J. Ware, J. Photogr. Sci., 1986, 34, 13.
12. A. Howie, Faraday Discuss., 1991, 92, 1.
13. M. Faraday, Phil. Trans. Roy. Soc., 1857, 147, 145.
14. M. Kerker, 'The Scattering of Light and Other Electromagnetic Radiation', Academic Press, New York, 1969.
15. G. Mie, Ann. Physik, 1908, 25, 377.
16. R. Gans, Ann. Physik, 1915, 47, 270.
17. M. Kerker, J. Colloid Interface Sci., 1985, 105, 297.
18. J.A. Creighton and D.G. Eadon, J. Chem. Soc. Faraday Trans., 1991, 87, 3881.
19. A. Henglein, P. Mulvaney, and T. Linnert, Faraday Discuss., 1991, 92, 31.
20. A. Henglein, J. Phys. Chem., 1979, 83, 2209.
21. E.F.F. Heuser, 'The Chemistry of Cellulose', John Wiley and Sons, New York, 1944.
22. R.A. Young and R.H. Rowell (Eds.), 'Cellulose: Structure, Modification and Hydrolysis', John Wiley, New York, 1986.
23. 'Recent Advances in the Chemistry of Cellulose and Starch', Ed. J. Honeyman, Heywood, 1959.
24. T.H. James (Ed.), 'The Theory of the Photographic Process', Macmillan, New York, 1977.
25. L.F. Grantham, T.S. Elleman and D.S. Martin, J. Amer. Chem. Soc., 1955, 77, 2965.
26. D.R. Stranks, 'The Reaction Rates of Transitional Metal Complexes' in J. Lewis and G.G. Wilkins, (Eds.) 'Modern Coordination Chemistry', Interscience, New York, 1960.
27. L.B. Hunt, Gold Bull., 1976, 9, 134.
28. P. Ellis, Gold Bull., 1975, 8, 7.
29. W.F. Berg, Gold Bull., 1979, 12, 97.

30. E. Bordignon, L. Cattalini, G. Natile, and A. Scatturin, J. Chem. Soc. Chem. Comm., 1973, 878; G. Natile, E. Bordignon and L. Cattalini, Inorg. Chem., 1976, 15, 246.
31. C.A. McAuliffe, R.V. Parish and P.D. Randall, J. Chem. Soc. Dalton Trans., 1979, 1730.
32. G. Frens, Nature, 1973, 241, 20.
33. J. Turkevich, P.C. Stevenson and J. Hillier, Discuss. Faraday Soc., 1951, 11, 55.
34. G.C. Britton, Inorganic Syntheses, 1978, 18, 201.
35. M.J. Ware, Brit. J. Photogr., 1991, 139, 17.
36. E. Wiegel, Z. Physik, 1954, 136, 642.
37. C.R. Berry and D.C. Skillman, J. Photogr. Sci., 1969, 17, 145; G.C. Farnell and L.R. Solman, J. Photogr. Sci., 1970, 18, 94.

Subject Index

Ablation, 47ff, 54ff
Absorption
  chemical, 227, 230, 233
Accelerated weathering tester, 182
Acrylate, 70, 81
Acrylic monomers, 32-35
  Acticryl CL959R, 42
  carbamates, 33, 38, 40
  carbonates, 33, 38, 40
  dioxolane, 38
  halogenated, 38
  multifunctional, 33, 38, 40
  oxazolidone, 33, 38, 40
  silicon/tin containing, 16
  urethaneic, 36
Actinometer, 254
Activation energy, 64, 79, 96, 205, 238, 242
Adhesion, 13
Adhesives, 1, 40
Alignment
  liquid crystals, 1
Amine synergists
  in photocure, 3, 26-29
Amino acids
  asparagine, 147
  aspartic acid, 147
  lysine, 146
p-Aminophenol, 229
Annealing, 76, 77
Anthocyanine dyes, 253
Antifoggants, 227
Antioxidants, 222
Archival permanence, 257
Argentotype emulsions, 262
Ascorbic acid, 227, 229
Asymmetric induction, 166
ATP generation, 145
Autocatalysis, 223, 227, 230, 231
  rate law, 224, 232
Azobenzene
  4-amino-4'-sulphone-derivatives, 134ff

Bacteriorhodopsin, 145ff
  photochemistry, 146ff
  photophysics, 146ff
Bathochromic shift, 158, 163, 175
Benzotriazole, 227
Bipolaron, 120, 125
Bleaching, 241-248
  photoinduced, 217
Blue print, 253
Bromide, 241
Butvar B-76R, 211-215

Calculations
  quantum mechanical, 138, 170, 201, 206, 222
Catalysis, 227, 263
Cationic cure, 4, 11, 209ff
Cellulose, 256
  quasi-solution in, 256
Chain transfer, 41
Charge barrier effect, 226, 232
Charge generating layer, 197
Charge transfer
  intermolecular, 110
Charge transport, 125, 197, 198
  layer, 197, 198
Chemical sensors, 6
Chemical stabilisers
  hindered amine, 181
  nickel complexes, 181
2-Chloro-4-aminophenol, 227
Chlorodiane blue, 197, 198, 204
Chrysotype emulsions, 253, 259
CIDEP, 28
CIDNP, 28
Cloud point, 81

Coatings, 1, 38, 39, 42, 209
Colas process, 253
Colloidal metals, 254, 257, 261-3
Colour images
  direct positive full, 220, 261
Colour negatives, 236-251
Colour-tuning, 129
Computational chemistry, 201, 206
Computer aided design, 7
Computer aided machining, 7
Conformational change, 69, 74
Contact printing, 257
Cooperativity
  in main chain polymers, 139-142
Corrosion, 223
Couplers, 240, 245
Covering power, 263
Critical size, 232
Crystallisation
  in polymers, 68
  thermally induced, 112
  vapour-induced, 112
CR-39 ^{R} monomer, 182
Cure
  cationic, 4, 11, 209
  electron beam, 16
  monitoring, 32ff, 68, 80
  shrinkage, 8
  thermal, 42
  UV (see UV curing)
  visible light, 3
Cyanines, 212
  borate ion bleaching, 217

Degenerate four wave mixing, 112
Depolarisation, 136
Developer agents, 222, 225, 227
  colour, 227, 229, 245
  concentration, 224
  negative charge, 224
  selectivity, 226
Developing
  colour, 236-251
  continuation phase, 224-232
  direct phase, 223, 232
  initiation phase, 224, 226, 230, 231
  process
    C-22^R, 238-242, 245
    C-41^R, 238, 242-251
    C-41B^R, 246
    C-41RA^R, 246, 248
    E-4^R, 242
  selectivity, 225

solution physical, 223, 227, 229, 232, 233
temperature, 238, 239, 242
time, 236-251
Diarylethene heterocyclics, 170
Diaryliodonium salts, 4, 209
Di-n-butylphthalate oil, 211, 215, 218
Dichromate printing, 253
Differential photocalorimetry, 29, 36
Diffraction efficiency, 105, 149
Diffusion
  oxygen, 74
  polymer chain, 74, 81
Dilatometry, 35, 36

Ektachrome^R, 242
Elastomers, 39
Electrical poling, 135
Electroluminescence, 120
Electron acceptors
  in photoinitiation, 20
Electron-beam curing, 16
Electronic energy migration, 96
Electron transfer, 3
  inner sphere, 229, 233, 259
  in ion pair, 216
  outer sphere, 229, 232, 259
  photo-induced, 216
Electro-optic devices, 120
Electrophotography, 194
Emulsions
  chrysotype, 253, 259
  palladiotype, 253, 254
  photographic, 225
End capped oligomers, 33
Enhancement factor, 141
Epoxy
  monomers, 4, 21, 23, 36
  resin, 68
EPR spectroscopy, 20
Etch depth, 49, 55, 59
Evanescent wave induced
  fluorescence (EWIFS), 98
Excimer
  emission, 69, 74, 76, 88ff
  laser, 47ff, 54ff, 106
Exciton, 120
  formation, 123
Eyewear
  photochromic, 179
  plastic lens, 179

Fatigue
  photoinduced, 145, 156, 170, 181
Ferricyanide, 241, 246, 253

Subject Index

Ferrogallate process, 253
Films
 laser iradiated, 55
 polymer, 54, 70, 102, 111
 stretched, 77
Fixing, 241, 243-248
 stability, 242
 time, 247, 248
Flexicolor, 237
Fluorescence
 depolarisation, 92
  spectra, 71ff, 87
 polydiacetylenes, 99
 polystyrene, 71ff, 88ff
 quenching, 216
 time-resolved, 71, 76, 87ff
Fluorimetry, 36
Fog
 developer, 223, 227
 emulsion, 223
Forehardening, 243, 245
Formaldehyde, 240
Fourier transform correlator, 150, 152
Free radical
 polymerisation, 1ff, 15ff, 32ff, 83
Fulgides, 156ff
 dicyanomethylene derivatives, 163
Fulgimides, 159
 chiral, 166
Fulminating gold, 261

Gallic acid, 253
Gamma radiation, 20
Gel
 effect, 41
 polystyrene, 69, 72
Gelatine, 211, 215, 218, 238, 248, 258, 261
 hardening, 240-245
Gelation, 80
Gel permeation chromatography, 91
Gibbs-Thomson equation, 225
Glass transition temperature, 68, 112, 205
Gold, 255
 complexes, 260
 prints, 253, 257, 259, 263
Grating
 laser-induced, 104

Hardening, 242, 243, 245
Head-up displays (HUDs), 9
Helix coil, 69, 76
Hexacyanoferrate(III), 241, 246, 253

Hologram
 real-time, 104, 148, 152
Holographic dry processable film, 10
Holographic optical elements (HOEs), 8-11
Holography
 of purple membrane, 147, 150
 Fourier, 150
 pattern recognition, 150, 152
Hydrazone DEH, 197
Hydrogen abstraction, 2
Hydroquinone, 229
Hyperpolarisabilities, 104, 134, 138, 142
 tensor, 135, 140
Hypsochromic shift, 147, 161

Imbibition
 with photochromics, 182
INDO calculations, 138
Induction period, 217, 223, 226
Infrared spectroscopy, 71, 199
 real time, 26, 32ff
Instant image, 169
Instant photography, 223
Integrated circuit manufacture, 1
Integrated optics, 134
Iodonium cations, 4, 209-219
Ionic displacements, 136
Irgacure [R], 15, 28, 29, 33, 44
Iridium, 254
Iron
 carboxylates, 253-265
 citrate, 253, 262
 EDTA, 245-248
 oxalate, 53, 254, 257, 258, 262
 1,3-PDTA, 246, 253
Iron-arene complexes
 as photo-initiators, 15, 23-25
Iron printing, 262
Isofulgimides, 159-163

Kallitype, 262
Kendall-Pelz rule, 222
Kodacolor[R] film, 238, 239

Langmuir-Blodgett films
 photochromic, 159
 nonlinear optical, 135, 142, 143
Laser
 ablation of polymers, 47ff, 54ff
 carbon dioxide, 54, 59
 curing, 32
 diode, 106, 194

Laser (continued)
  excimer, 47ff, 54ff
  induced grating, 104
  interferometry, 36
  krypton ion, 33
  in optical recording, 147ff, 156, 166, 169, 170
  power density, 49
  pulsed, 36, 41
  ultraviolet, 47, 54
Latent image, 222-227, 257
Lead salts,
  in platinotype, 258
Light emitting diodes, 120

Main chain polymers
  linking units in, 139
  in NLO, 134ff
Mass spectrometry, 226
Mercury,
  in photographic prints, 254, 257
Merocyanines, 212
Microcrystallites, 112
Microdensitometer test, 211
Mixed stack crystals, 117
MNDO treatment
  of photochromics, 170
Molecular dipoles, 135
MOPAC calculations, 138, 211, 219
Movement tracking, 153
Multiphoton processes, 102

Network, 68
5-Nitrobenzimidazole,
  as antifoggant, 227
Noble metals, 252ff
Noncentrosymmetric structure, 135
  Y-type L-B film, 142
Non-destructive readout, 148, 156, 166, 170, 175
Nonlinearity
  near-resonance, 111
  resonance enhanced, 111
Nonlinear optics, 102ff, 110ff
  second order effects, 134ff
  third order effects, 102, 110

Onium salts, 4
Opthalmic lenses, 179ff
Optical density
  solvent/binder, 215
  time response, 213, 218, 224, 228
Optical fibres, 6
Optical information processing, 145

Optically detected magnetic
  resonance (ODMR), 127
Optical nonlinearities
  third order, 102, 110
Optical recording
  fatigue in, 156, 166, 169, 170
  holographic, 147
Optical storage media, 156, 166, 169
Order parameters
  in L-B films, 142, 143
Organo-borate anions, 217
Orientational relaxation, 137
tris-Oxalalatoferrate(III), 254
Oxidative imaging dye
  bleaching, 212, 217
Oxonol dyes, 209ff
Oxonol-iodonium salts, 209ff
Oxygen
  inhibition, 18, 82
  quenching, 74
Oxygen sensitivity
  in epoxy cure, 23
  in photochromics, 172, 181
  in polymerisations, 18, 23, 33, 39

Palladiotype emulsions, 253, 254
Palladium, 253, 254, 257, 258, 263
Phase morphology, 84
Phase separation, 84
Phase transitions
  polystyrene gels, 81
p-Phenylenediamine, 227-230
Phenylhydrazine, 229
Photoablation
  carbon dioxide laser, 54ff
  excimer laser, 47ff, 54ff
  in waveguide formation, 106
Photoacoustic transients, 60
Photochromic materials, 145ff, 156ff, 169ff, 179ff
  fatigue resistant, 156, 167, 170, 171, 181
  infrared active, 156, 165, 174
  liquid crystals, 159
  optically active, 156, 166
  polymers, 169ff
  proteins, 145ff
Photochromic reaction
  cycling of, 147, 172, 177, 178
Photoconductivity, 194ff, 222
Photoconductors, 194ff
  fatigue, 194
Photocrosslinking, 10, 32, 81
Photocure
  UV (*see* UV curing)
  visible light, 1, 3

Subject Index

Photocyclisation, 199
Photocycloaddition, 91
Photodegradation, 145, 159, 180
Photodissociation
  of fulgides, 166
Photo-DSC, 29, 36
Photographic speeds, 225, 226
Photography
  fast, 51
Photoinitiators, 1-5, 15ff, 82
  Irgacure™ 15, 28, 29, 33, 44
  iron-arene complexes, 15, 23-25
  type I, 28, 29
  type II, 26, 29
Photolink[R] technology, 12
Photolithography, 15, 32, 42
Photoluminescence, 120
Photomerocyanine, 8, 180, 186, 189
Photometric equivalent, 263
Photoresists, 91
  negative, 16
  deep UV, 15ff
Photosensitisers, 18ff
Photovoltaic energy sources, 1
Physical aging
  of polymer, 137
Piezoelectric sensor, 51
Plasmons, 255
Platinotype, 253, 254, 257
Platinum prints, 252-254, 257, 258, 263
Plume
  photoablation, 51
Polarisation holography, 150
Polaron, 120
Poled structures
  stability of, 139
Polyacrylamide, 147
Polyalkylmethacrylate
  spacers in L-B films, 142
Polydiacetylenes, 98, 103, 110
Polyetheretherketone (PEEK), 55ff
  carbon fibre composites, 58
Polyethyleneterephthalate (PET), 55, 59, 197
Polymer
  composites, 58, 110
  surface, 47ff, 98
Polymerisation
  ring opening, 8
  solid state, 42
Polymers
  conjugated, 98, 102, 120
  from CR-39[R], 182

Polymethylacrylate
  copolymers, 93
Polymethylmethacrylate (PMMA)
  composites, 111
  copolymers, 93
  photoablation, 49
  resist, 16
Polyparaphenylenediacrylic
  acid (PPDA), 91
Polyparaphenylene-ethylene, 103
Polyphenylenevinylene (PPV), 121
Polysaccharides, 258
Polystyrene, 87, 177
  atactic, 72
  isotactic, 68ff
Polythiophenes, 103
Polyvinylalcohol, 147
Polyvinylchloride, 42
Polyvinylnaphthalene
  copolymers, 93ff
Printing
  dichromate, 253
  iron, 262
Print out, 256, 257, 262
Print processing
  business, 236
  machinery, 236, 246
Proton pump, 145
Pulse radiolysis, 225
Pump-probe, 117
Purple of Cassius, 259
Purple membrane, 145 ff
Push-pull chromophores, 134

Quantum chemical calculations, 138, 170, 201, 206, 222
Quantum confinement, 111, 255
Quantum yield
  for bleaching, 156, 158
  for byproduct formation, 172
  for colouring, 157, 158, 180
  for initiation of polymerisation, 41
  for photocure, 35, 41
  for photodegradation, 181
  for photoinitiation, 44
Quenching
  fluorescence, 216
  oxygen, 74
  thermal, 76
  triplet state, 29
Quinolines, 25
Quinoxalines
  reaction with oxygen, 15, 25-28

Radiation cured coating, 209
Radical pair, 20, 216

Real-time optical correlators, 152
Real-time processing, 148
Real-time UV spectroscopy, 35, 44
Refractive index change, 156, 148
Relaxation motions, 135-138
Reptation, 81
Reticulate doping
  polymers, 109
Retinal protein, 145ff
Rheology, 67
Rheometer, 82
Riboflavin, 25
Rose Bengal, 216, 217
Rotational correlation times, 93
Ruthenium, 254

SCF calculations, 201
Second-harmonic
  signal, 136, 143
  phase-matched, 143
Segmental motion, 95
Semiconductors
  doped glasses, 118
  one-dimensional, 110, 120
  polymers, 102
  II-VI, 118
Sepiaprint, 262
Side-chain polymers
  in NLO, 135
Silanes, 16-23
Silver
  (0), 222ff
  colour, 263
  clusters, 225
  halides
    emulsions, 222ff
    surface charge, 227
  recovery, 242, 246
  sulphamate, 262
Single photon timing, 71
Sizers, 256, 258
Solid state polymerisation, 42
Solvatochromism, 98
Space charge, 136
Spatial resolution, 149, 169
Spectral response
  analysis test rig, 212
Spironapthoxazines, 181ff
  synthesis, 181
Spiropyrans, 179
Stannanes, 16-23
Stereolithography, 7, 32
Stopped flow kinetics, 223-229, 238

Surface
  free energy, 225
  modification, 1, 12
  reactivity, 231
  wettability, 13

Tack-free cure, 18, 23
Tannic acid, 253
Tergitol TMN-10^R, 211
Tetrachloroaurate(III), 259
Tetracyano-p-quinodimethane
  (TCNQ), 110ff
Tetrachloroplatinate(II) anion, 253
Thermally stimulated discharge
  current (TSDC), 136
Thiocyanate, 248
Thionorbornene, 8
Thiosulphate, 241, 243, 246, 247, 262
Third harmonic generation, 104, 110
Threshold fluence, 56, 58, 62, 64
Triplet state quenchers
  in photocure, 29

UVA light, 254, 257
UV curing
  cationic, 4, 23-25, 35, 44
  epoxy resin, 68ff
  free radical, 1-4, 15ff, 35, 38, 41

Van Dyke process, 262
Vinyl ethers, 4
Viscosity, 68, 82
Visual pigment, 145

Waveguide
  attenuation, 143
  from L-B Films, 143
  optical, 106, 118
Wetting agent, 241

X-ray analysis
  of fulgides, 167
  of hydrazone DEH, 207
  of spiroxazines, 191

Z-scan, 112, 117